The Rewiring of America
The Fiber Optics Revolution

The Rewiring of America
The Fiber Optics Revolution

C. David Chaffee

Atlantic Information Services
Washington, DC

ACADEMIC PRESS, INC.

Harcourt Brace Jovanovich, Publishers

Boston   Orlando   San Diego
New York   Austin   London   Sydney
Tokyo   Toronto

ACADEMIC PRESS, INC.
Orlando, Florida 32887

United Kingdom Edition published by
ACADEMIC PRESS, INC. (LONDON) LTD.
24-28 Oval Road, London NW1 7DX

Library of Congress Cataloging-in-Publication Data
Chaffee, C. David.
    The rewiring of America.
    Includes index.
    1. Optical communications.   2. Fiber optics.
I. Title.
    TK5103.59.C465   1988   621.38′0414   87-18709
    ISBN 0-12-166360-4

88 89 90 91   9 8 7 6 5 4 3 2 1
Printed in the United States of America

To Margot and Catie,
Hubert and Charlotte

Contents

Prologue

To the men of Madison Lightwave Cable Company, fiber optics is not some pie-in-the-sky, otherworldly technology. It is how they make their living.

Tonight, June 26, 1986, they are installing optical fiber cable in duct space leased from C&P Telephone Company. Their job is to run more than 5000 meters of the cable through the underground ducts made available to them. The cable is to run from Bradley Boulevard in Bethesda, Maryland, to Tenley Circle in Washington, D.C.

The work crew started at 6 P.M. on this Thursday evening, as the rush of rush hour was dying and the heat was beginning to subside.

Jamie McCoy and Jim Wood slowly slide the cable off the big reel that is supplied by Northern Telecom from Canada. The cable snakes down into the manhole, where it disappears. (See Figure 1.)

Crew foreman Linwood Snow stops by, fresh from one of the puller locations up the road. There are the usual jokes about how things go better when the owner isn't around, who is being lazy and who isn't, etc.

The operation began when an air compressor literally blew a balloon with a string attached to it through the duct and to the first puller station, which is several thousand yards away. At the puller site, the balloon is detached and the string is wound around a pulley as it comes up through the puller site's manhole.

Figure 1. Huge reels can hold miles of thin optical fiber cable. (Photo courtesy US Sprint.)

Attached to the string is the optical fiber cable, which starts being fed off the reel after the string has gone completely through the duct joining the manhole where Wood and McCoy are and the first puller station. It is hard to believe that the black cable, which is less than one-half inch thick, contains 50 optical fibers.

When the man stationed at the first puller site uses the two-way radio to announce that the cable has been sent through, Wood and McCoy stop reeling it off the wheel.

The air compressor, towed by a truck, stops at the first puller station. The men get the hose out and blow the balloon, with string attached, up through the duct to the second station. The cable follows behind and is reeled to that point.

The procedure continues through the night until the optical fiber cable reaches the seventh station at Tenley Circle in Washington. Throughout the process, two C&P inspectors—one on the Maryland side and one on the District side—have been monitoring the activity.

The crew, which totals 14 men, are held up near the end as the string breaks several times. When that happens, the string must be relocated by scampering down into the sewer to retrieve it and then using the compressor gun to send the balloon and string along the way again.

"The sewers aren't bad," explains C&P inspector Ken Smith. "There aren't any rats, although there are cockroaches—some big ones, too."

It is after 1 A.M. as the cable finally reaches its destination at Tenley Circle. The men are used to working these hours, from 6 P.M. to 6 A.M., but they are starting to get tired.

Figure 2. Hundreds of work crews have participated in the rewiring of America in all kinds of terrain and weather. (Photo courtesy US Sprint.)

When the cable reaches Tenley Circle, the crew still has to go back to where McCoy and Wood are. From there, the remaining cable is wound into a figure eight. The next step is to move the cable up Bradley Boulevard the other way, where it will be strung along utility poles owned by the local power company, Pepco.

In order to continue, Snow must summon a splicing crew. When asked if the splicers don't mind being awakened in the middle of the night, Snow laughs: "They're probably out drinking beer somewhere."

Snow waits patiently for the splicing crew to get there. The job will be done before he goes to bed, around dawn. Snow has been working 60-hour weeks for some time now and will be glad when this job is done. He hopes the next assignment for installing the optical fiber cable will be out in the countryside, perhaps in the Midwest where he is from.

The Madison team is one of hundreds of crews throughout the United States whose job it is to install optical fiber cable. It is all part of a process that will revolutionize the way we communicate with one another in ways that were once thought to be impossible.

The Madison team trudges off at the end of the job, knowing only that it is on call to soon do exactly the same thing somewhere nearby. Its job tonight represents but one small link in what is becoming the rewiring of America. (See Figure 2.)

Chapter 1

Birth of a Technology

To understand the birth and rise of fiber optics is to gain a glimpse into the best of what mankind has become. Its story is one of human struggle, of corporate infighting, of countries racing to best one another. But it is also the story of cooperation, of realizing the importance of contributing to something that is larger than just an individual, a company, or even a country.

It is the story of the growth of a new technology whose scope is nothing less than the revolution of how we communicate with one another, whose mission is to provide us with ways of interacting that before had only been dreamed of.

It is the story of trying and failing, but more importantly of trying. For while there have been winners and losers in this struggle, all were guided at least to some extent by the principle that it is laudatory to leave the world a better place than one has found it. The implementation of this new phenomenon—fiber optics—will certainly do that.

Anyone who has seen sunlight stream through the clouds and come in through a window in a single ray is familiar with optical energy. Optical communications is simply the attempt to use light and the photons that constitute light to transmit messages. Smoke signals are a primitive example of optical communications.

1

In 1880, Alexander Graham Bell created the photophone, which used sunlight to transmit spoken sounds. Bell intoned: "I have heard a ray of sun laugh and cough and sing. I have been able to hear a shadow, and I have even perceived by ear the passing of a cloud across the sun's disk."[1]

Bell's discovery may have served as an inspiration to the work to follow at Bell Laboratories, but it ignored one basic fact: human beings were not going to communicate optically by harnessing light traveling at 186,000 miles per second that came from the sun. In order to communicate using light, people were going to have to create their own source of light. They were also going to have to figure out their own means of transmitting the light and then design their own way of decoding it so people could understand the message being sent.

This source of light did not come until the late 1950s, with creation in the United States of the laser (acronym for Light Amplification by Stimulated Emission of Radiation). A laser could produce light in pulses, but it was not sunlight. And the laser used power that was electrically generated.

AT&T's Bell Laboratories played an instrumental role in developing the laser. Bell Labs was also credited with creating the first semiconductor chip in the 1940s. However, it was parrallel independent work done at three other labs (General Electric, IBM, and MIT Lincoln Labs) that led to the announcement of the first pulsed semiconductor lasers, which operated at very low temperatures. Bell Labs, however, was to later create the first units operating at room temperature. (See Figure 1.)

"As the laser appeared, I and my associates at Bell Laboratories saw that there was in the laser as a coherent source of light-wave energy the possibility of transmitting tremendous amounts of information and that it was a valuable transmission medium," explains Stewart Miller, who helped pioneer the technology at Bell Labs.

---

[1]*Semiconductors and Semimetals*, Vol. 22, Lightwave Communications Technology, Academic Press, Inc., Harcourt Brace Jovanovich, Publishers, 1985.

Figure 1. Fiber optics would not be possible without the invention of the laser. One variation of semiconductor lasers pulses millions of times per second through optical fiber to transmit messages. (Photo courtesy AT&T Bell Laboratories.)

While the laser was the driver, the source, and the transmitter, a separate essential element was the receiver, a device necessary to accept the signals transmitted from the laser and then decode them into a form in which they could be used to transmit something practical, such as a voice message or stream of data.

The basic types of receivers, known as photodiodes, had actually been in existence since the 1950s, prior to the birth of the laser. In the 1950s, "there were semiconductor detectors, there was just nothing much to do with them," Miller remembers. "They were used in research work in physics—as detectors for photons."

The interest in optical communications right after the advent of the laser was intense. "We were committed to optical communications even before the fibers," Miller says. "We started right after lasers appeared and fibers were one of several types of transmission media we were exploring, but we had the commitment to optical communications before that. We had a major effort before any other group in the world."

As scientists became interested in optical communications in the next several years following introduction of the laser, a fundamental question arose: Could signals be transmitted from these lasers to detectors by using the atmosphere, as with smoke signals, or would it be necessary to control the atmosphere in which they were delivered?

These researchers realized that the open atmosphere was not the most reliable medium for the transmission of light. Clouds, fog, and obstructions saw to that. So it became necessary to develop conduits to send light from one place to another. There was much talk about what the conduits should consist of and, as any successful technology evolves, many more would-be technologies were discarded in the search for the right medium.

One early maker of these fibers was Elias Snitzer, who did research at the American Optical Company in Southbridge, Massachusetts, in the early 1960s. Snitzer and a few associates published a paper on the propagation characteristics of optical fiber suggesting that fiber could be used as a communications medium. Snitzer even went so far as to make optical fibers for carrying transmission signals, although they were only a few inches long.

The scene shifted to the United Kingdom in the mid-1960s where work was going on to develop the necessary controlled atmosphere. Important research was taking place at Standard Telecommunications Laboratories (STL), which prompted work to be initiated at British Telecommunications, then known as the British Post Office.

Probably the most cited paper in the short history of fiber optics was put together by Charles Kao and G. A. Hockham. Kao was heading up a research team at STL at the time, and the

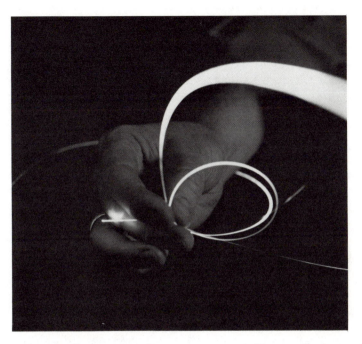

Figure 2. Early researchers found that optical fiber made from glass was an excellent medium for transmitting laser light. It was small, lightweight, and easily carried photons. The trick was to manufacture fiber that could carry the light at loss levels low enough to make it economical. (Photo courtesy AT&T Bell Laboratories.)

paper was drawn from three years of research. Entitled "Dielectric-fiber surface waveguides for optical frequencies,"[2] it suggested that thin pieces of glass could be used to send laser signals at a signal loss of 20 decibels per kilometer (dB/km) or lower. When the glass— or fiber—was good enough to send signals at that level, that's when things began to get "interesting," according to Kao. (See Figure 2.)

The paper caused a stir in the scientific community for several reasons. For one thing, sending a light pulse that distance with

[2]"Dielectric-fiber surface waveguides for optical frequencies," by K. C. Kao and G. A. Hockham (1966). *Proc. IEEE* 133, 1151.

those losses meant that you could equal the transmission capabilities of the conventional metal cables that were commercially being used. For another thing, it suggested that the existing levels of loss when glass was used (which were now up in the tens of thousands of decibels per kilometer) could theoretically be brought down to an acceptable level for communications purposes. Kao suggested that glass could be used to reach the magical 20 dB/km loss level.

Meanwhile, a technical liaison for Corning Glass Works, Dr. Gail Smith, visited the British Post Office. While there, Smith was told of the need to develop pure glass fibers for communications and was encouraged to bring the challenge back to Corning, which was known for its understanding of glass.

It was now clear that the basic elements of what constituted a fiber optic link had been defined. These included the source of light (the laser), the optical fiber, and the receiver. Yet to say this is all it takes to transmit voice or data is an oversimplification.

First of all, a person's voice is transmitted through an electronic switching circuit to program a light source to send the voice message. So the message is transmitted from an electronic medium to an optical medium, from electrons to photons. The associated hardware that makes this transition possible is known as opto-electronics.

The optical, or photonic, signal is carried through the transmission medium, the optical fiber, until it reaches the receiver, which decodes the signal back electronically (to electrons) where it is amplified for the listener.

The electronics in a fiber optic system consists of the transmitter, receiver, and associated opto-electronics. This is differentiated from the cable, which simply carries the signal.

Each fiber has a core for transmitting the laser's light and a cladding. Both the core and cladding are usually made from glass. The cladding acts as a wall to keep the light contained in the core. The laser light travels differently, depending on how the core and cladding refract the light, and the core and cladding have different refractive indices.

Figure 3. Even before fiber had reached the low losses that would make it suitable for communications purposes, it was fused or tied together in bundles for imaging purposes. One application was for medicine where, inserted in the body, fiber could help to image internal organs. (Photo courtesy AT&T Bell Laboratories.)

The intensity of the light pulse as it is traveling through the fiber inevitably gets weaker, due to a variety of factors. The loss of light is known as attenuation. At the time of the Kao paper, the attenuation level was very high. The challenge was to make it very low.

That does not mean that the early high-loss optical fibers were totally worthless. In fact, the first applications of fiber optics were for purposes of imaging rather than communications. The idea was to take thousands of high-loss fibers, bind or fuse them together, and then transmit light through them to create an image. (See Figure 3.)

Medicine represented one early application. By using a 3-foot bundle of fiber, you could gain an image of just about every part of the human body.

Still the challenge that had been posited in the British laboratories related to low-loss fiber, and it was taken seriously in the United States both at Corning and at Bell Laboratories. As a result, work began in earnest. If it was possible to get losses down to 20 dB/km, researchers wanted to find the way. But, as in any new trip into the unknown, these same researchers were unsure whether the process would ever really work, or whether this technology would be like the hundreds of others that for some reason had fallen by the wayside.

"We didn't know if you could ever reach those levels," explains Peter Schultz, one of three members of the original Corning team assembled to try to crack the 20 dB/km barrier. The team also included Robert Maurer and Don Keck.

"It was like looking into a cloud and you didn't know what was really in there, how transparent could glass be? There was no way to say it could be that transparent, no evidence that it could really be that transparent," Schultz remembers.

An immediate problem in trying to get glass so pure that it could send laser signals for long distances was that there was no established means, no test equipment, to figure out how to get the impurities to the parts-per-billion level required. This represented a "significant challenge," Dr. Kao says. "Here, one has material, one has the theory, but one can't prove by measurement that everything that was predicted was true because the measurement techniques were way off. It was like taking a meter stick with 10 centimeter calibrations and trying to measure millimeters." Still, Kao observes, "the human race has been fairly good at being able to achieve many things without being able to precisely define them."

But Corning did not let the problem of not knowing how to measure what they were attempting to create get in the way of its research. "We knew things were going to be so pure that we had a difficult time measuring the chemical purity and our real figure of merit was to measure the optical transmission of the fiber," explains Schultz. In other words, Corning would create a fiber and keep testing it empirically until its ability to send laser signals longer distances got better.

When you are creating a technology, and this was not the first time Bell Labs was doing that, you learn to create everything you have to in order to get the job done, explains Miller. "You don't ask anybody how to do it, you do it yourself. We originated the techniques for measuring the various losses for fibers for the first time, as well as how to get the low losses."

Nor, Miller acknowledges, did the researchers know what other obstacles might present themselves along the way, including some basic law of science that the researchers might have been ignoring. "As it turned out, nature smiled," he says.

The AT&T process, known as the modified chemical vapor deposition process, begins with the manufacture of a high-purity glass tube (See Figure 4.), which is then cooled (See Figure 5.) The tube is then slowly dropped from a drawing tower where it is heated in a furnace and then drawn into fibers (See Figure 6.) The resulting fibers are then cooled and characterized in assembly-line fashion. (See Figure 7.) From there, they are cabled and sent out to the job. (See Figure 8.)

There was competition in the air, particularly between Corning and Bell Labs, as each raced to get the loss down to the 20 dB/km level. Yet it was a curious interworking, one not easily defined.

When Schultz is asked whether Corning and AT&T were arch rivals, friendly competitors, or allies in promoting a new technology, he responds "All of the above, I guess. It was in the best interest of both to try to promote the technology." Yet he notes that there was "competitiveness" and "some rivalry." Still, "There was also interaction, although it was a little at arm's length, but there was interaction between lab people and between business people."

It was this interaction that allowed fiber optics to go forth as it did. "There was a symbiotic relationship between the two groups which had an expertise that meshed," Miller says. "We knew what it took to make a system, they knew glass to make experimental fibers and we worked on a technical basis back and forth. We got some experimental fibers from them and carried out system experiments with them."

The whole thing came wonderfully together by 1970. Maurer, Schultz and Keck had reached the magical level of 20 dB/km in their optical fiber experiments.

Corning's technique for manufacturing optical fiber at the time was the vapor deposition process, which was different from AT&T's modified chemical vapor deposition process. As the

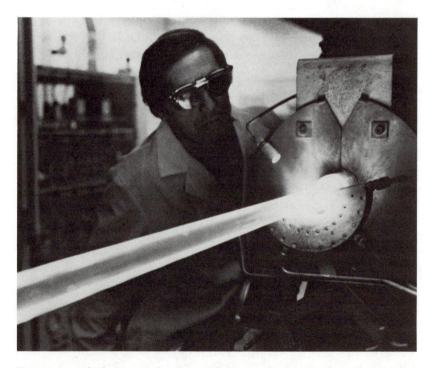

Figure 4. A lathe is used to manufacture special glass preforms. The operator in this case is AT&T's Harry Dzikowski. (Photo courtesy AT&T.)

Figure 5. A glass preform just prior to its extrusion into optical fibers using AT&T's modified chemical vapor deposition process. (Photo courtesy AT&T.)

result of a cross-licensing agreement signed in 1970 involving electronic ceramic materials, the two companies each agreed to let the other use their optical fiber manufacturing process.

In 1970, Hayashi and Panish at Bell Laboratories demonstrated the first continuous-wave, room-temperature-operated semiconductor laser.[3] The potentials of small size, high reliabil-

---

[3]"Junction lasers which operate continuously at room temperature," by I. Hayashi, M. B. Panish, P. W. Foy, and S. Sumski (1970). *Applied Physics Letters,* 17, 109.

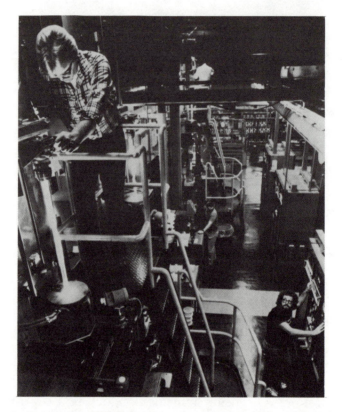

Figure 6. Using the AT&T process, the preform is then placed in the fixture at the top of the drawing machine *(above left)*. The lower end of the preform is then slowly lowered into a furnace, which heats the glass until it is molten. The softened tip is pulled down through the machine to form a continuous flexible fiber. (Photo courtesy AT&T.)

Figure 7. Once the fiber is drawn, it is characterized in assembly-line fashion. Here AT&T cable plant employee Pat Jones operates one of many electronic testing machines used to check the performance characteristics of optical fibers before they are ready for market. (Photo courtesy AT&T.)

Figure 8. The finished form—AT&T optical fiber on a spool ready for cabling. (Photo courtesy AT&T.)

ity, low loss, long life, and the ability to reach very high rates of transmission loomed ahead.

The technology for receivers was already in place. The elements of a basic functional fiber optic system were defined and working in the laboratory. But before those developing these products could even think of coming to market, there was still much work and refinement to be done.

Chapter 2

Fiber Optics: First Steps

NOW THAT A technology known as fiber optics had been created, the founders began searching for the refinements that would eventually bring it to the marketplace.

It became evident right after the successful semiconductor laser experiment that the laser had a long way to go before it would reach commercial viability. "They ran continuous wave for 10 minutes and that was the end of it," Miller remembers about the first units in 1970.

Bell Labs began seeing the advantages of an alternative to lasers, which came in the form of the light emitting diode (called LED). LEDs are similar to lasers in that both convert electrical signals into optical pulses and both are generally made from the same components. LEDs are more stable than lasers, although the area from which they emit light is much broader and less focused than is that of a laser, and the resulting light is not as powerful. Lasers are more directional, providing rays of light that are more coherent.

At any rate, the LED's stability—its ability to remain relatively immune from such factors as heat, humidity, and age—put it in good stead with those who were refining fiber optics and hoping to bring it rapidly to market.

Researchers were also trying to determine what the best

wavelength was at which to operate a laser or LED. "In the early days it started at 633 nm, then 700 nm, then 800 nm, and finally 850–840 nm," Schultz remembers. And the 800–850 nm range is where it was to stay for some time.

But it was this decision to go ahead with LEDs and not fully attempt to refine and utilize the advantages of lasers that caused the leading practitioners to diverge from using the initially developed single-mode fibers researched by Snitzer, Kao, AT&T, and the Corning team and to originate and refine mult-imode fibers. The reasoning at the time seemed simple and practical. First, coupling LEDs with multi-mode fiber "could create a communication facility," explained Miller. Multi-mode fiber had a much larger inner core than single-mode fiber (although in laymen's terms both were tiny, literally little larger than the size of a human hair) and could thus more easily be coupled with an LED or laser than could single-mode fiber. (See Figure 1.)

Miller, who earlier had advised Corning to manufacture single-mode fiber, now requested that Corning make multi-mode fiber for AT&T. "We never lost interest in single-mode, the greater emphasis on multi-mode for a while was purely from the fact that LEDs were practical, inexpensive, and reliable, and the lasers had not yet achieved that."

To back up Miller's contention that the Labs maintained an avid interest in single-mode fiber, an early paper presented by a group of fellow Bell Labs researchers headed by Leonard Cohen noted that single-mode fiber "may be considered as the super-highway of future optical communication systems."[1]

Corning's response to Miller's request was to develop a new type of multi-mode fiber beyond the step-index multi-mode fiber it had worked on, which transmitted light in a zigzag path.

Trying to achieve the distances inherent in single-mode design and the easy coupling of step-index multi-mode fiber,

---

[1]"Transmission Characteristics of Large-Core, Low-Loss, Single-Mode Fibers," by L. G. Cohen, P. Kaiser, and H. M. Presby (1975), presented at the first Optical Fiber Communications conference in Williamsburg, Virginia.

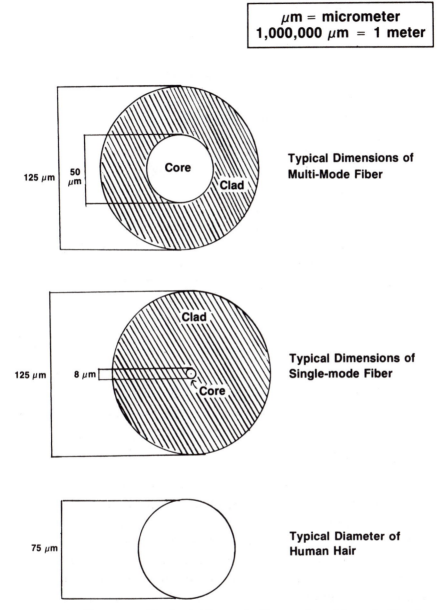

μm = micrometer
1,000,000 μm = 1 meter

125 μm   50 μm   Core   Clad   Typical Dimensions of Multi-Mode Fiber

125 μm   8 μm   Clad   Core   Typical Dimensions of Single-mode Fiber

75 μm   Typical Diameter of Human Hair

Figure 1. The core of both single-mode fiber and multi-mode fiber is smaller than the diameter of a typical human hair (which is approximately 75 microns). However, with the cladding that forms the outer diameter, both are slightly larger than a human hair. (Diagram courtesy of Orionics.)

Corning went one step further and developed graded-index multi-mode fiber, through which paths of light proceed in waves. The core of a graded-index multi-mode fiber is radially graded in composition, in such a way that the path length differences between refracting light rays is minimized. The result was that light loss was substantially reduced from what step-index fibers could produce. (See Figure 2.)

While at that time it may have made sense to take the LED-multi-mode fiber path, we will see in later chapters that the decision to use LEDs and multi-mode fiber generally and multi-mode fiber in particular may have resulted in the United States losing important technological ground and bringing systems to market that were obsolete not long after they were installed. On the other hand, however, sticking with what must have been considered the more exotic laser–single-mode combination would probably have delayed the advent of the first commercial systems.

Kao characterizes research on multi-mode fibers as "unfortunately diverting our attention quite a bit during the development stages." He continues: "We could have stuck to our guns and said that single-mode fiber was the way to go and is the best solution and in that case we wouldn't have all these problems that we solved. It turned out that they were unnecessary problems."

Kao remembers being one of the few who was not ecstatic about graded-index multi-mode fiber when it was the talk of the first major conference held specifically on fiber optics. Kao says he explained to others at the conference that only single-mode fiber could provide the extra-long distances, but said those comments obviously did nothing to stop the further development of graded-index multi-mode fiber. A significant reason for that was that research was being dominated by Corning and Bell Laboratories; despite its early efforts, ITT simply did not have the base at that time to challenge those organizations.

Yet the development of multi-mode fiber raised for the first time the hope that fiber optics could actually be used in the com-

Multimode Fiber

**Refractive
Index Profile**

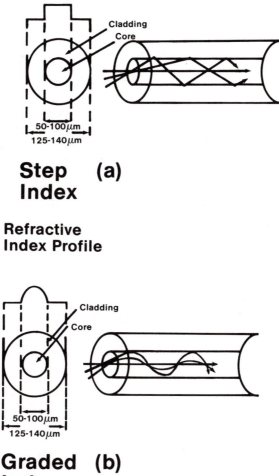

Step **(a)**
Index

**Refractive
Index Profile**

Graded **(b)**
Index

Figure 2. Schematic illustration of the propagation of light rays in the multimode (*a*) step- and (*b*) graded-index fibers. In step-index fibers, the rays may be thought to follow a zigzag path, while in graded-index fibers the paths are transmitted in waves. (Graphic courtesy Corning Glass Works.)

mercial marketplace. A plethora of other issues relating to the development of fiber optics was also being addressed successfully. Coatings were being developed to minimize the cracking that glass otherwise might have been subjected to; they also provided greater flexibility. Strength characteristics were being built into the fiber. Fiber drawing, the process by which fiber is extruded from the glass, was also becoming a rapidly improving art.

Other aspects that were being addressed included splicing two fibers together and holding them together with connectors. It was impossible for Corning or Bell Labs to precisely draw the length of optical fiber used for each job. For one thing, that drawing length was limited, usually to less than 2 km. As a result, researchers had to develop ways of splicing two fibers together into one.

Schultz characterized the development of splicing techniques as coming about more easily than he first had thought. "I was surprised that you could take two fibers, stick them together, heat them up, fuse them together, and have such good coupling," he said. (See Figure 3.)

The researchers favored silica-based glass as the transmission medium of choice. "Silica is one of the simpler glasses that achieved historic low losses and is a system that has good imaging," Kao explained. In the early stages, techniques for making silica glass were also more advanced than other techniques.

One great feature of pure-silica fiber was that it came from sand, which represented an inexhaustible resource. It was doped with germanium, which came in much more limited supply, but this did not pose an insurmountable problem.

Corning was continuing to register lower attenuation losses with its fiber, down from 20 dB/km to 10 dB/km, to 5 dB/km, and so on. Schultz notes that in 1972, when Corning had reduced the loss to 4 dB/km, "that's when we realized we really had something." As a result, Corning did marketing studies "and realized these fibers could be quite attractive, quite competitive." (See Figure 4.)

Figure 3. The fine art of splicing involved joining two optical fibers together in the field, often using a repair vehicle. Here a Northern Telecom splicing machine is used to join two fibers together. (Photo courtesy Northern Telecom.)

"There was a recognition that it was going to be a very important communications type medium," recalls Miller. "As a matter of fact, as early as 1973, I and two coauthors, Enrique Marcatili and Tingye Li, wrote a review paper[2] in which we clearly outlined the broad applicability of fiber transmission in the intercity and transoceanic markets. It was clear it was coming as early as 1973."

The research effort took on a new direction as engineers began preparing the first commercial prototypes in 1974. AT&T was attempting to manufacture lasers at the Murray Hill, New Jersey,

[2]"Research toward optical fiber transmission systems," by S. E. Miller, E. A. J. Marcatali, and T. Li, *Proceedings IEEE* 61, 1703–1751 (December 1973).

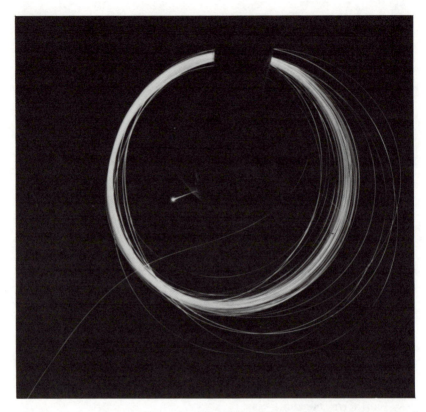

Figure 4. Corning continued to decrease the loss levels of the optical fiber it was developing. By 1972, it began to realize that fiber could be competitive in the communications marketplace. (Photo courtesy AT&T.)

facility and optical fiber at its Product Engineering Control facility in Atlanta.

While using LEDs had been such a selling point for the development of multi-mode fibers, Bell Labs engineers were once again recognizing the advantages of the laser as they scaled up for the first prototype. The problem with the LEDs was chromatic dispersion, "which completely kills you" at the shorter wavelengths that were being used, explains Bell Labs engineer Ira Jacobs, who was directly involved. One could only go several kilometers with LEDs without needing to get the signal recharged, but one could reach seven kilometers with lasers.

But if the Labs decided that it could go with lasers, why not revert back to single-mode fiber? "You could make a single-mode fiber," Jacobs notes, "the problem was in coupling light into it, of connecting it, of splicing it." Jacobs continues: "In the period of the early 70s, those were considered very formidable problems for multi-mode fiber, and the single-mode fiber was an order of magnitude more difficult. At that stage, single-mode fiber was purely a research kind of concept, although it was recognized, even at that time, that the ability of single-mode fiber was greater than that of multi-mode fiber."

Miller, together with Jarus Quinn, executive director of the Optical Society of America, was the driving force behind the first fiber optics conference held in Williamsburg, Virginia, in 1975. That conference has arguably become the most important forum on fiber optics in the world and has played an essential role in helping to acquaint outsiders with the technology. Start up of the conference was in itself an important indicator of how far the technology had come.

But 1975 was also important because it marked the end of the incubation period for fiber optics. After leading a sheltered early life in the laboratories, it was time to present the technology to the outside world. By December, Bell Labs announced to its employees that such a system was in the making. The day of bringing fiber optics to market was at hand.

Chapter 3

The Battle for Acceptance

THE FIRST PROTOTYPE consisted of 2100 feet of optical fiber cable that was installed in ducts and manholes adjacent to the Bell Labs Product Engineering Center in Atlanta. Because a large number of fibers was installed between the two areas, repeater-less signals were actually sent distances of up to 7 km.

While the optical fiber was multi-mode, each of the sources was a handpicked laser from the New Jersey facilities. LEDs were not employed.

The lasers pulsed 44.7 million times per second (44.7 Mbps) and were the driving forces behind AT&T's new FT3C series electronics, which corresponds to the third level of the digital hierarchy used in America. While Bell Labs had successfully tested one and two series digital electronics, which operate at approximately 1.5 Mbps and 6 Mbps, respectively, the problem was that the economics were not as good for transmitting smaller amounts of information.

The first cable carried a surprisingly large number of fibers (144) and was advertised by AT&T as having the capability of carrying 50,000 telephone calls. The 144 fibers actually consisted of 12 fiber ribbons, each ribbon containing 12 fibers. The Labs claimed the ribbon idea was a good one, because it allowed them to splice 12 fibers together at a time via a mass splice, rather than having to splice one fiber at a time. (See Figure 1.) Both AT&T and Corning provided optical fiber for the experiment.

AT&T also introduced another new feature, "pigtails," which hang at the end of the fiber for easy coupling with the transmitters or photodetectors and "eliminate the need for special handling of the fibers," according to Bell Labs.

"The first Bell System applications for lightwave communications systems are likely to be in carrying information digitally

Figure 1. An optical fiber cable, similar to the one shown here, carried voice, video, and data signals in the Atlanta experiment. The cable, protected with steel wires embedded in its sheath, contains 12 ribbons, each encapsulating 12 fibers. (Photo courtesy AT&T Bell.)

between telephone switching centers in metropolitan areas," said Bell Labs, in introducing the system. "In such areas, space in underground cable ducts is limited and digital transmission already is used extensively."

The Atlanta experiment was helping AT&T to prepare for "possible widespread use by telephone companies in the early 1980s, and perhaps even sooner for special applications," Bell Labs announced. "Ultimately, lightwave communications systems may be used throughout the entire telecommunications network."

Selling features of fiber optics that the Labs enumerated in that early announcement included potential low cost, immunity from electromagnetic interference, the ability to carry large amounts of data in "noisy environments," and the potential for homes or businesses to employ video applications using the fiber.

By the end of 1976, Bell Labs was calling the experiment a success. "The technical feasibility of lightwave communications has been proven—in an environment simulating that of metropolitan central offices," wrote Ira Jacobs.[1]

The word was beginning to get out about fiber optics. One favorable article first appearing in *Barron's* in 1976[2] was later excerpted in *Science Digest*[3] under the banner "Fiber Optics: At the Threshold of a Communications Revolution." The article noted that fiber optics can carry great amounts of information and did not suffer from cross talk, as did its competitor, copper wire. It also said that the U.S. military was interested in fiber optics, spending about $35 million annually.

It seemed appropriate that AT&T would follow its Atlanta trial up with the first commercial fiber optic system, but there was a usurper waiting in the wings. On April 25, 1977, GTE Corpora-

[1]"Lightwave Communications Passes its First Test," by Ira Jacobs, *Bell Laboratories Record*, December 1976.

[2]"Shedding Fresh Light (Optical Communications May Be the Wave of the Future)", Barron's, Oct. 11, 1976, p. 3.

[3]"Fiber Optics at the Threshold of a Communications Revolution," by Jim Powell, *Science Digest*, February 1977.

tion announced "the world's first optical communications system to provide regular telephone service to the public."[4]

The 5.6-mile California link carried traffic between the General Telephone Company of California long-distance switching center at Long Beach and its local exchange building in Artesia. A scant 24 telephone calls were carried over the link, which included six fibers.

GTE Laboratories obviously had been researching fiber optics on its own. GTE took the occasion of the link's announcement to sell aggressively the wonders of fiber optics to the outside world. "Optical transmission systems are expected to bring many benefits both to the public and the telephone industry through their ability to handle, on an economical basis, more than a thousand times the number of phone calls and other communications signals handled by existing copper wire systems," explained Lee Davenport, president of GTE Laboratories.

GTE used its own LEDs and multi-mode fiber provided by Corning Glass Works for its prototype. It promised the link would remain operational for at least one year. GTE particularly ballyhooed the pulse code modulation scheme it employed. The PCM systems were used to convert conventional analog voice signals into digital signals and were developed by GTE Lenkurt, a GTE manufacturing subsidiary.

GTE explained its network in the following way. As customers' analog voice signals entered the Long Beach switching center, they were converted into binary digits and assigned to individual "time slots" for transmission at the other end. The high-speed streams of electrical pulses were fed into the "miniscule light emitting diode," according to GTE, which then transformed them into pulses of light that traveled through the optical fiber. The signals were sent 1.6 miles, where they were boosted by a repeater. A second repeater rebooted the signals another 2.3 miles down the road.

From there, the light signals entered the Artesia switching

---

[4]GTE press release issued April 25, 1977, 11 A.M.

center 1.7 miles from the second repeater, where they were again converted by a photodetector into electrical pulses that were fed into PCM equipment, which decoded them and reconstructed the customers' voices. The voices then traveled through the local telephone lines to their destination in Artesia.

GTE also announced that it was developing a system to send television signals 1300 feet to provide color television pictures and that voice and voice/video could be carried over two fibers. GTE said it had implemented new splicing techniques and used a new type of cleaving tool.

The GTE revelation provided a boost in the marketplace that may have been helpful in that it showed there could be competition. Corning, for one, was not going to get very far if it had to wait for AT&T to provide it with orders. And Corning at that time, Stewart Miller notes, was trying to sell fiber wherever it could.

Less than three weeks after the GTE announcement, a chastened AT&T announced "the world's first lightwave communications system to provide a wide range of telecommunications services."

This 1.5-mile network was in Chicago, sending signals between two Illinois Bell switching offices and between one of those facilities and a downtown Chicago office building housing a number of customers. Unlike the Atlanta experiment, the Chicago experiment used only two 12-fiber ribbons. Four optical fiber breaks were found. Also unlike the Atlanta experiment, both lasers and LEDs were used.

Both GTE and AT&T went on to announce further links in almost rapid-fire sequence.

In June 1977, GTE and the Hawaiian Telephone Company, a GTE subsidiary, installed the first fiber optic link in Hawaii. The short link connected two data processing centers at Camp H.M. Smith, headquarters of the U.S. Pacific Command.

AT&T in 1978 installed optical fiber cable on telephone poles in the snow at Bell Labs' Chester, New Jersey facility, to see how environmentally sound the cable was. Initial measurements, according to Bell Labs, demonstrated no signal loss differences from underground installations.

In March 1979, GTE subsidiary General Telephone Company of Indiana installed a fiber optic link for connecting two telephone switching centers in Fort Wayne, Indiana. In the same month, British Columbia Telephone Company, a GTE subsidiary, activated the first fiber optic network in that Canadian province, a 4.5-mile link connecting switching centers in Vancouver. The Fort Wayne link was supplied by GTE Lenkurt and was characterized as being one of the first that would be permanent.

"The new optical communications system will cost General Telephone of Indiana about $300,000 to install initially and will represent an investment of $2.8 million after additional terminal facilities have been placed into service to bring the link to full capacity," said Andrew Buckner, Vice President, Network Engineering and Construction. "The use of optical fiber technology on this project represents an approximate 36% saving from the estimated $4.3 million, which would have been required to install a conventional system utilizing copper cable," according to Buckner.

GTE was not afraid to sell the advantages of fiber optics, noting the far greater capacity offered, the range of services possible, and "significantly lower installation costs." Other advantages included smaller size—allowing road crews in some cases to install it in existing duct space rather than tearing up streets—and immunity from cross talk, electromagnetic interference, and lightning.

Soon after the Fort Wayne link was up and running, this last advantage—immunity from lightning—was driven home starkly. A number of engineers and craftspeople were at the plant one summer night in 1979 when a severe thunderstorm went through Fort Wayne. The copper T-1 carrier system was knocked out. The fiber optic network kept running smoothly. "The craft people at the plant immediately saw the advantages of fiber optics," remembers Lowell Krandell, who was a transmission engineer on the project.

Bell Labs helped Southern New England Telephone Company install a link in April 1979. It stretched between a telephone company switching office and customers' homes.

GTE provided the first fiber optic network in Belgium beginning in July 1979, a 6.5-mile trunk linking Brussels and the nearby suburb of Vilvoorde.

While GTE was introducing fiber optic links to other countries, it was clear that the rest of the world had not been oblivious to fiber optics. The Japanese had begun their own prototype fiber optic experiment, Hi-Ovis (Highly Interactive Optical Visual Information System), which linked 158 homes outside Osaka, Japan, with fiber optics. The project allowed villagers to get as many television channels as they wanted, shop by television, tap into train timetables, or carry out a host of other functions.

Charles Kao acknowledges that he played a role in bringing the technology to the Japanese. But, with regard to Hi-Ovis, he explains that it in some ways had a negative impact. "It is good that fiber can be brought into everybody's home, but it was a rather expensive way of doing it, so Hi-Ovis has its role in history, but not in promoting fiber optics."

The Japanese were causing some heads to turn on another front, however. NEC announced in 1977 that it had developed an experimental 400 Mbps digital transmission system using selfoc fiber, which is made of multicomponent glass.

The Canadians separately were announcing in 1979 that they had installed the first major fiber optic cable television trunk in North America. The field trial was taking place in London, Ontario.

In October 1979, Bell Labs began a field experiment in Sacramento, California, connecting Pacific Telephone central switching offices.

The following year AT&T introduced fiber optics to the Olympics. This was one of its few analog fiber optic links, used for transmitting television signals and voice communications in Lake Placid, New York.

In April 1980, New York Telephone operated a fiber optic link allowing telephone company computers in lower Manhattan to talk to each other.

But the real breakthrough announcement of 1980 was that AT&T intended to build an uninterrupted 611-mile fiber optic network between Cambridge, Massachusetts, and Washington, D.C. This long-distance application not only meant that fiber was going to be competitive with metal cable, it also meant that fiber optics was going to take on a much newer and much more exciting technology—satellite.

It also meant that AT&T had enough faith in fiber optics to assign perhaps its most populous trunk to the new and budding technology. If people in the communications industry hadn't heard about fiber optics, or weren't taking it seriously, the so-called Northeast Corridor intended to change that in a hurry.

Chapter 4

Fiber Scores Its First Breakthroughs

THE STORY OF the Northeast Corridor is not simply the story of the successful fulfillment of the first major fiber optic network operating in the United States, even though that is probably what AT&T intended. It is also the story of a monopoly favoring its own subsidiary, of a schizophrenic Federal Communications Commission giving AT&T contradictory marching orders, and of the devastating realization that the Japanese were winning a significant part of the game.

When AT&T told the FCC in early 1980 that it intended to build the Northeast Corridor network, it was confirming its own confidence in fiber optics as a technology and as a means of carrying significant portions of its own traffic.

Patiently employing the scientific method and relying on 20 years of intensive study, AT&T was ready to begin making true the promise of fiber optics.

AT&T's initial filing for the Northeast Corridor project in 1980 proposed that Western Electric would build the network. The route was to be divided into two parts: the first phase, between New York and Washington, D.C., and the second phase, between New York and Cambridge, Massachusetts. Cost through 1984 was estimated to be $79 million.

AT&T at first declared that its FT3C electronics, together with multi-mode fiber, would be employed. This was the same electronics that had been used in AT&T's digital voice experiments and was consistent with the decision earlier to use multi-mode fiber at the 800 nm wavelength. AT&T would upgrade the electronics to 90 Mbps some time after it had begun operating.

AT&T was staying with its earlier predilection of having large amounts of fiber in the cable, with cable containing from 48 to 144 fibers. Repeater spacings were to be set at approximately every four miles, rather than the one-mile spacings AT&T would have used with coaxial cable.

While in the 1970s it had been GTE Corporation that had given AT&T a run for its money, the competitor this time was MCI Telecommunications Corporation.

MCI demanded that the FCC deny AT&T's request, at first arguing that AT&T had not addressed a variety of economic concerns, including cost of right-of-way and price of retiring analog switches, and demonstrating that fiber optics would not soon become obsolete. They demanded that AT&T correctly factor in the research costs that had been associated with fiber. In a later filing, MCI argued that AT&T had not done a thorough evaluation of other fiber optic suppliers before revealing that its own subsidiary, Western Electric, would be its supplier.

In making Western Electric the sole supplier, AT&T also summarily dismissed Corning—its old co-research pioneer and friendly competitor—from the job. In its comments, Corning contended that if Western was indeed the only supplier, then Western would automatically emerge as the low-cost bidder, forcing competitors from the marketplace.

By not allowing competition, Corning argued that it would be impossible to determine whether the optical fiber could have been made at a lower cost. It also noted that foreclosing the Bell System market to other manufacturers would diminish the necessary funding potential competitors required for the levels of research and development necessary to improve the technology.

Corning proposed that the FCC adopt a policy of competitive procurement for the Northeast Corridor, where specific lot sizes of optical fiber would be held out for bid. This would ensure the development of a healthy industry and result in lower prices.

Corning criticized the multi-mode fiber proposed by AT&T, claiming it could produce single-mode optical fiber operating at longer wavelengths (1300 nm) with much lower transmission losses.

While AT&T argued that competitive bidding was impractical, it soon became apparent from the tone of letters from the FCC's Common Carrier Bureau that the FCC felt there was some merit to the request to open the bidding. A staff letter issued to AT&T on June 27, 1980 requested that AT&T do a comparative analysis on the cost of using optical fiber cable produced by Western Electric, as opposed to using fiber manufactured by other suppliers.

While allowing AT&T to go forth with Western Electric as its supplier for the New York–Washington network, the commission demanded that AT&T open to competition the second phase of the network—that running between New York and Cambridge.

Realizing that the FCC would otherwise delay—or deny altogether—its Northeast Corridor project application, AT&T informed the FCC on September 19, 1980 that it would issue a solicitation seeking competitive bids for the second phase of the network. In that correspondence, AT&T also claimed that certain new technological innovations would allow it to use 90 Mbps from the beginning, rather than having to upgrade to 90 Mbps.

In the order it adopted November 25, the FCC denied MCI's petition and allowed AT&T to build the first phase of the Northeast Corridor project using Western Electric as its supplier. While noting that it would take significant growth to fill the circuits proposed, the FCC believed AT&T's argument that its expenditures would be justifiable in the long run. The FCC reasoned that allowing AT&T to go ahead with the New York–Washington network would allow for lessons learned there to be addressed in the second phase.

While concerned about the closed procurement in the first phase, the FCC said some of its worries had been alleviated by the September 19 letter from AT&T promising open procurement for the second phase. "We believe that a competitive market will result in increased innovation and, ultimately, in reduced prices to the consumers," the FCC said.

From the sound of it, the FCC was also beginning to see the advantages of fiber optics. It said: "End-to-end digital communication will be made more cost effective using this technology [fiber optics], which may revolutionize the capabilities of the domestic communications network."[1]

AT&T issued a request for information on March 4, 1981 to 108 companies that might be interested in providing equipment for the New York–Cambridge leg of the Northeast Corridor. Of those, 43 responded that they were interested. On April 21, a solicitation was sent to those companies. Specific proposals were then submitted by Western Electric, the French company Citcom Systems, Inc., Harris Corporation, North America Philips/CSD, and four Japanese companies (Hitachi America, Ltd., NEC America, Inc., Sumitomo Electric U.S.A., and Fujitsu America, Inc.).

As the Japanese solicitations were unveiled, it became clear that—while AT&T had been rolling out its prototype networks operating at 45 Mbps—the Japanese had been developing much faster fiber optic electronics and more advanced systems generally. Companies such as NEC and Fujitsu were working on electronics in the 400 Mbps range in the late 1970s, literally ten times faster than the electronics AT&T had developed.

But how did Japanese companies reach this level of technology, while pioneer AT&T had not? Charles Kao characterizes the

---

[1]FCC Memorandum Opinion, Order and Authorization (File No. W-P-C 3071), adopted November 25, 1980; released January 27, 1981.

Japanese work "as a remarkable bit of engineering to get the processes involved under control," and says that the Japanese "meticulously studied the peripheral technologies."

AT&T's Stewart Miller is less charitable, noting that the work represented "a short-range differential in efforts." He continues: "The only thing that is to be noted with regard to those efforts is that they [the Japanese] have a very large number of people that are highly trained and devoted to this area. And as a result of their culture, they have a need to export and their whole economy is based on doing things for export. They focused on optical communications as one area they specialized in and put a big effort into that effort."

The all-encompassing research drive of the Japanese was a key factor in their taking the lead. "They have the ability to put a large number of people into all parallel possibilities in order to explore them all—not to be able to prejudge by selection—but to do them all," Miller says. "That has been significant."

The Fujitsu bid stood out as the strongest of those submitted by the Japanese companies. It was the lowest bid and represented the most advanced technology. By the end of 1981, Fujitsu claimed that it had already provided a variety of transmission systems to a number of entities.

Fujitsu claimed that it could offer transmission systems operating at 800, 400, 100, 45, 32, 6.3, and 2 Mbps. [The 800 Mbps claim was an exaggeration, those electronics were not commercially used until five years later.]

Fujitsu claimed its customers had included Nippon Telegraph and Telephone Public Corporation, Hong Kong Telephone Company, the Taiwan Ministry of Communications, Singapore Telecom, United Telephone Company of Florida, Western Union, and Satellite Business Systems. [The reason why companies such as Corning did not respond to the solicitation was because AT&T demanded that only those suppliers that could provide a turnkey system should apply. Corning was a fiber supplier, but it was not involved with electronics.]

In its proposal, Fujitsu said it would use Siecor Corporation to be its optical fiber cable manufacturer. A joint venture between the West German-based Siemens and Corning, Siecor would no doubt have used large amounts of Corning fiber.

Fujitsu proposed implementing 135 Mbps technology, which permits the same capacity as Western Electric's proposed 90 Mbps—with approximately one-third fewer fibers and repeaters. The Fujitsu proposal would also have employed 1300 nm wavelengths, longer wavelengths than AT&T proposed, resulting in lower loss factors and a substantial reduction in the number of repeaters. Fujitsu calculated that it would cost $33.3 million for it to supply the Cambridge–New York segment.

AT&T and New England Telephone officials toured Fujitsu's plants in Japan between August 19–22, 1981, a tour that Fujitsu representatives said was entirely successful. At a follow-up meeting on September 8 including AT&T, Fujitsu, and Siecor representatives, "Fujitsu was given every reason to believe that it had submitted the lowest and best bid for the project and that AT&T was inclined to accept its proposal," according to a later Fujitsu filing with the FCC.[2]

Word that AT&T was going to award the second phase of the Northeast Corridor project to Fujitsu got out. Republican South Carolina Senator Strom Thurmond wrote the White House a memo dated September 16, 1981, explaining that a Japanese firm had offered the lowest bid for the second phase of the Northeast Corridor and noting that such a bid "would establish a foreign corporation as the essential operators of our sensitive communications network on which our Defense Department would be dependent."

Several weeks later, on October 16, Representative Timothy Wirth (Democrat-Colorado), Chairman of the House Subcommittee on Telecommunications, Consumer Protection and Finance, wrote FCC Chairman Mark Fowler a letter. Wirth

---

[2]Petition to Deny by Fujitsu America, Inc. (W-P-C 3071, W-P-C 4173) Filed December 11, 1981.

brought up national security concerns and questioned whether the FCC should "open up a promising technology for internal network equipment to a foreign bidder."

Three days later, Senator Daniel Inouye (Democrat-Hawaii) expressed his "grave concern" to Fowler that "this contract may be awarded to a foreign-led consortium and that a decision favoring foreign electronics is imminent. . . ."

Responding to reports of Fujitsu's success, one of the five FCC commissioners, Joseph Fogarty, recognized the FCC's culpability. In a memorandum to his fellow commissioners dated October 21, Fogarty said: "The Commission must bear heavy responsibility for this bad result because we literally forced AT&T to enter into procurement procedures, the only possible result of which could have been the awarding of the contract to a Japanese firm such as Fujitsu . . . The question before the Commission now is how can we make sure that AT&T, due to our past pressure, is not forced to award this major contract to a foreign firm."

The following day, October 22, Senators Robert Packwood (Republican-Oregon), Barry Goldwater (Republican-Arizona), and Harrison Schmidt (Republican-New Mexico) urged the FCC to consult with the Defense Department and amend the order to specify that AT&T was not required to accept the lowest bid.

By October 29, the FCC had done a complete about-face and told AT&T to address national concern issues. On that date, AT&T awarded the contracts for the second phase of the Northeast Corridor project to Western Electric, in compliance with the FCC—the same FCC that earlier had told AT&T it must have open procurement for that route. In making the announcement, AT&T attributed the decision in part to concerns "raised in government circles about so critical a communications artery being provided by a foreign vendor."

Exacerbating the situation to some extent were charges by Commissioner Fogarty and others that Fujitsu had "low-balled" its way into the contract. The claim was that the other Japanese suppliers had set Fujitsu up to win the contract by providing higher bids and generally making their offers less attractive. Here Fujitsu demanded, and received, an apology from Fogarty.

Fujitsu's reaction to the whole business—understandably— was bitter. "The barrage of charges of illegality against Fujitsu has caused substantial and wholly unwarranted injury to Fujitsu's reputation, not only in the United States but in Japan, where the accusations were widely reported and where charges of improper or illegal corporate conduct are viewed with abhorrence," Fujitsu told the FCC.

"The unfortunate irony of Fujitsu's position in this entire affair is inescapable," Fujitsu told the FCC. "Invited by AT&T to participate in a fair, open competitive procurement, Fujitsu invested substantial amounts of time and money in the preparation of a technically faultless proposal." The final result was that Fujitsu lost the contracts and "found itself faced with serious, albeit unsubstantiated, charges of illegal conduct."

Fujitsu acknowledged that it was "unclear" whether it was pressure from the government that made AT&T reverse its position or whether these pressures simply gave AT&T the pretext to go back to its earlier position of only allowing Western Electric to supply the equipment.

"What is clear," Fujitsu said, "is that AT&T all too gladly accepted the fruits of a well-organized exercise in xenophobia, and that the net result of this sad episode is that AT&T has been able to avoid the result of its own procurement process. . . ."[3]

This may not be entirely fair to AT&T, which was conscientiously attempting to comply with the orders of an admittedly mixed-up Federal Communications Commission. An argument not heavily debated was that AT&T and other U.S. suppliers would not receive the same consideration in Japan were the shoe on the other foot, although Fogarty did mention it. The U.S. government certainly did not act in a consistent manner, but was that worse than a consistent Japanese government that system-

[3]Ibid.

atically closed its market to foreign suppliers and which gave its suppliers an edge by funding their R&D activities?

Miller believes strongly that there is something to the argument that the Japanese government provides its companies with an unfair advantage. "Their government has promoted a cooperation and collaboration in business in an evolutionary sense for companies," he says, "whereas ours' has been out with a big stick looking for any place where there might be an antitrust inference. And that has had a dampening impact."

There is also some credence to the Defense Department's claim that national security was a factor. While it probably would have been more advantageous for the Department of Defense (DoD) to have made its concerns known earlier, the Pentagon did have a valid interest.

In its FCC filing, made after AT&T had already decided to exclude foreign suppliers, DoD noted that—through its Defense Communications Agency—it leased long-haul telecommunications at an annual cost of more than $500 million, nearly $300 million of which was paid to Bell System Companies. "The DoD thus has an important national security interest, and a significant consumer interest, in the development, manner of deployment, and cost of new telecommunications technology such as fiber optic cables," DoD told the FCC.[4]

DoD had come to see the advantages of using fiber optics because fiber optics significantly decreased the vulnerability of U.S. telecommunications resources to disruption by electromagnetic pulse. DoD also noted that domestic R&D would have been discouraged had Fujitsu received the job. "From DoD's perspective, it is critical that this nation's defense industrial base be expanded, and the effect of AT&T's proposal to award the contracts to a domestic supplier will stimulate the U.S. R&D base."

It should be pointed out here that the Pentagon has been, and continues to be, a major customer of AT&T. DoD, for example,

---

[4]Comments of the Department of Defense (File Nos. W-P-C 3071 and W-P-C 4173), filed with the FCC December 11, 1981.

announced in 1985 that it was awarding AT&T a contract to service its Defense Commercialization Telecommunications Network, a contract worth hundreds of millions of dollars. The Northeast Corridor is being used as part of that network.

In response to the FCC's concern that it wanted more competition in the supplying of optical fiber for phase two of the Northeast Corridor, AT&T announced that it would be purchasing 100 megameters (100 million meters) from outside suppliers. Receiving portions of the job were Corning, ITT Telecommunications, and Valtec Corp. The DoD supported this action in its comments.

But this concession benefited Corning more than is first realized. In addition to being named one of the three suppliers, Corning also sued both ITT and Valtec for patent infringement, making the other two pay royalty fees to Corning for the fiber they produced. Corning also brought suit against the U.S. government for buying optical fiber from other suppliers, which, Corning argued, it should have received royalty payments for. All three suits were eventually decided in Corning's favor.

Such was the pose that Corning would continually strike when it came to protecting legally the optical fiber that it had so diligently worked to develop from the beginning. While we will study the impact of this position, there are several ironies. Certainly the fact that Dr. Kao's company, ITT, had to pay royalties to Corning is interesting, since Kao was among those who created the technology in the first place. (For the record, Schultz steadfastly maintains that Corning's entrance into fiber optics had nothing to do with the Kao paper.) It is also ironic that Corning, who went to the FCC to complain that AT&T was not opening the optical fiber cable market to outside suppliers, moved to shore up its interest against other would-be competitors as soon as that assurance was granted, in large measure creating a two-supplier market.

Still, Corning's argument that opening the bidding for fiber would advance the technology had some merit. Had AT&T used the single-mode fiber that Corning said it could provide, it would not have had to come back years later and overlay the route with single-mode fiber.

In December 1982, AT&T revealed that it planned to extend the Northeast Corridor from Washington, D.C., to Moseley, Virginia. Western Electric was again to be the supplier.

It was in that same month that MCI began to let it be known that it was interested in fiber optics as well. MCI announced that it was leasing right-of-way from Amtrak to install a fiber optic network between Washington, D.C., and New York City. The 20-year lease agreement allowed MCI to increase its circuit capacity to 40,000 along that corridor, four times the amount that its terrestrial microwave system was delivering.

With this announcement, MCI provided a bravado that got people's attention. President Orville Wright explained that MCI's fiber optic network was going to be better than AT&T's because it would be single-mode fiber based. MCI's approach to fiber optics was more spontaneous than was AT&T's. Already sensing that the coming divestiture of AT&T was creating a whole new ball game, MCI was in a hurry to gain marketplace credibility.

This exciting new technology, fiber optics, gave MCI a chance to install a transmission medium that had unlimited capacity—something it could grow into. Fiber optics was not cheap—no question about it—but word had gotten around in the financial community about MCI, and it was flush with cash. MCI also had the advantage of not being encumbered as a regulated carrier, unlike arch rival AT&T.

There was probably another reason why MCI was so excited about fiber, something that really got its corporate adrenaline going. That was that MCI could beat AT&T at its own game. For while AT&T had helped to create fiber optics and had doggedly sought to improve it for 20 years, AT&T was not the master of fiber optics technology—as the Northeast Corridor project had shown.

Then as 1983, or the last year before divestiture, arrived, MCI and AT&T made revealing statements that would detail how their futures would be carved out. MCI's announcement was destined to enliven the marketplace and push into second gear a technology that had already gotten its act on the road.

In early January, MCI announced that it was ordering large amounts of single-mode optical fiber cable from Siecor Corporation and Northern Telecom. Here was a carrier willing to pay tens of millions of dollars for tens of thousands of kilometers of a fiber that AT&T had shied away from using and that no one had yet to employ.

A major reason Siecor was selected was because it was Corning's major cabler; the two both had plants in North Carolina, although the relationship was not an exclusive one. Northern Telecom was chosen because of the large amounts of optical fiber cable it had produced for Canadian networks.

The Canadians were also finding out about fiber optics and were attempting to fit its benefits to Canada's rural population. In Elie, Manitoba, a field trial was set up to serve 150 households between the farming communities of Elie and St. Eustache, Manitoba. The $9.6 million trial was inaugurated on October 23, 1981; it provided a variety of services, including single-party digital telephone, cable TV, stereo, FM radio, and Telidon services, such as specialized information of interest to farmers, including grain and livestock prices.

But the even bigger news was that Saskatchewan Telephone Company was planning a $50 million network that would stretch 3200 km by the time it was completed. Pumping out the large amounts of optical fiber cable for that was Northern Telecom, which had built a plant in Saskatoon. Despite some early start-up problems, Northern Telecom had a good deal of experience by the time of the MCI announcement.

At about the same time MCI startled the budding fiber optics industry, AT&T announced that it was going to install five major fiber optic routes throughout the United States beyond the Northeast Corridor and use Western Electric as its supplier. This time Western Electric said it had developed high-speed electronics, to 432 Mbps, and was using single-mode fiber. AT&T was destined to stay with Western Electric as much as possible.

AT&T kept the ball rolling in February 1983, by announcing that the heart of its Northeast Corridor project—between New York and Washington—was now operating. The 372-mile network used approximately 30,000 miles of optical fiber. A video-conference for the benefit of the press employing the fiber optics was held at both ends. New York stock exchange quotations were sent to Washington, and pages of the *Congressional Record* were transmitted to New York.

Bell Labs President Ian Ross acknowledged at the event that the system could have had much greater capacity had single-mode fiber been used. Ross indicated that wavelength division multiplexing would eventually increase the 90 Mbps transmission rate to 270 Mbps.

At the Optical Fiber Communications seminar soon thereafter, AT&T announced another breakthrough. In one of its so-called "hero experiments," Bell Labs had sent an unrepeated message 119 km using single-mode fiber. Data rates of 420 Mbps were achieved.

The whole idea of hero experiments is something that cropped up in fiber optics as the technology was coming to market. The experiments are done in pure laboratory settings and cannot directly be related to actual commercial networks. Miller analogized their impact to the fiber optics industry to what high-speed car racing is to the automotive industry. Miller noted that hero experiments provide an incentive to the people developing them and can provide motivation for those interested in investing in fiber optics.

By the end of March, AT&T announced that it had turned on the first leg of its West Coast fiber optic network. This route joined Sacramento and San Jose and included Oakland, San Francisco and Stockton. The system, which used the same 90 Mbps technology as the Northeast Corridor, cost $38 million.

At about that time, AT&T announced that it had also operated a 7-mile commercial network in Louisiana that used long wavelength lasers and fiber and operated at the 1300 nm wavelength.

MCI announced in May that it picked Fujitsu to provide its fiber optic electronics between Washington, D.C., and New York City. There was admittedly something poetic about MCI's coupling with Fujitsu. MCI was, of course, playing David to AT&T's Goliath, fighting to be competitive in whatever way it could. And here was Fujitsu, spurned soon after it had been given assurances by AT&T that it would be one of its major suppliers, hooking up with MCI.

Fujitsu promised to provide electronics operating at 405 Mbps at the 1300 nm wavelength using single-mode fiber. Repeaters were to be spaced every 20 miles, compared to AT&T's 4-mile spacings along the same corridor. While AT&T said it would employ 432 Mbps electronics in the near future, Fujitsu was destined to beat it to the punch.

MCI also announced that it had signed a lease agreement with CSX Railroad, through which MCI would have access to 4000 miles of CSX right-of-way. MCI was to pay CSX $32 million for the 25-year lease. In return, MCI could string fiber optics throughout the eastern half of the United States, joining together such population centers as New Orleans, Atlanta, Miami, Philadelphia, Chicago, Detroit, Cleveland, Pittsburgh, Baltimore, Cincinnati, Indianapolis, St. Louis, Louisville, and Washington, D.C.

It appeared that AT&T and MCI were in for a real donnybrook and that outside suppliers would have to hover around MCI if they wanted action. However, there was an entire group of new carriers waiting in the wings that knew about the advantages of fiber optics and that were eager to build their own networks.

Chapter 5

The Economics of Fiber

HOW RAPIDLY FIBER optics came to the marketplace was a situation dependent on a variety of factors, each of which was weighed carefully by the communications engineers whose job it was to design successfully a carrier's network.

By 1980, fiber had long since passed the initial benchmark set by Kao and Hockham, 20 dB/km losses, which had been calculated as the level necessary to make it competitive with copper cable. Still, there were a number of other factors that had to be taken into account.

Certainly the strong commitment by AT&T—including Bell Laboratories and Western Electric—was an overwhelming factor in favor of developing fiber optics. The 22 Bell Operating Companies were part of AT&T and represented an incredibly strong captive market. The fact that AT&T believed in fiber optics had a very positive impact on the Bell Operating Companies, as well as the established independent telephone companies that worked with them and sometimes competed against them.

To exemplify AT&T's commitment to fiber optics, AT&T International's Art Schiller explained in 1983 that "fiber fever" was catching hold. "Fiber cable is being installed right now in backyards, along highways and railroad tracks, in office buildings, and in industrial parks and shopping malls," he said.[1]

AT&T chose fiber optics for its Northeast Corridor not as an experiment, but "because of the economies and capacities involved with up to 100,000 equivalent voice channels traversing the same cable," Schiller said. "The existing digital switching in this high-traffic route rendered the use of fiber optics a sound business decision."

Yet for all Schiller's cheerleading, fiber optics still had to be tailored to fit the right applications. The AT&T experience had shown that fiber optics was beginning to cost in when used for long distance voice and data applications. This was primarily because the reduced number of repeater stations helped to save money when compared to copper cable.

Another factor the communications design engineer had to take into account was capacity, and fiber optics could provide plenty of that. The question, in some instances, came down to long-term versus short-term planning. If there was significant traffic anticipated, then the engineer could grow into fiber optics. If much lower levels were expected, it made sense to upgrade existing systems.[2]

It was only in those applications in which carriers hoped to have as much capacity as possible that the higher speed systems that suppliers were developing became a factor. AT&T's Northeast Corridor project was providing significantly more capacity than alternate transmission media could provide, yet much greater capacity could be provided by using even better electronics.

The 405 Mbps electronics that Fujitsu could provide was more than four times the level that AT&T could supply. While Fujitsu

[1]"International Markets for Fiber Optic Communications," by Arthur Schiller, presented at Kessler Marketing Intelligence seminar in Newport, Rhode Island, 1983.

[2]"The Medium Is the Message," by C. David Chaffee, *Communications Systems Worlwide Magazine*, February 1985.

said it could deliver its superior electronics in 1983, and did, it took AT&T another two years to come to market with its higher speed equipment. AT&T had originally told the FCC it would employ 432 Mbps electronics in 1983; after revising the design, AT&T began installing 417 Mbps equipment in late 1985, providing Japanese suppliers an important window of opportunity to have the higher speed U.S. fiber optic electronics market to themselves.

The fact that AT&T had installed significant amounts of multi-mode fiber also eventually became a problem. The longer wavelength lasers were better suited to providing signals over single-mode fiber, as was mentioned earlier. To use the higher speed electronics meant that AT&T in some instances had to overlay single-mode fiber over the multi-mode fiber, which optimally operated at lower wavelengths over shorter repeaterless spans.

The early multi-mode systems featured large fiber counts, and each pair of fibers was forecasted to carry one or two DS-3 circuits, the equivalent of 672 or 1344 voice-grade circuits, according to John Bigham, a Bell of Pennsylvania- Diamond State Telephone engineer.

"After the development of single-mode fiber with higher bit rate multiplexers, this strategy was modified, making possible the use of lower fiber count cables and higher bit rate multiplexers. Using single-mode fiber and advances in multiplexers, the ability to increase the transmission speed increases the capacity for growth," Bigham told the 1986 Kessler Marketing Intelligence (KMI) conference.[3]

AT&T's philosophy in the early 1980s was that fiber best served the large metropolitan areas and that going to smaller towns was not economically justified. As the heat of competition to rewire as much of America as soon as possible was turned up by the mid-1980s, however, AT&T upgraded that strategy with

---

[3]"Deployment Strategies for Fiber in the Distribution Loop," by John Bigham, presented at the Kessler Marketing Intelligence conference in Newport, Rhode Island, 1986.

the emphasis on providing transcontinental fiber optic capability.

When AT&T announced that the first phase of its Northeast Corridor project had opened, it held a press conference, demonstrating that voice, video, and data could all be sent using the trunk. The ability to satisfy a variety of transmission requirements that fiber optics afforded was a major plus. Yet experiments for transmitting services over fiber optics for shorter applications were not evolving as rapidly.

Times-Fiber Communications, a major cable television supplier, had skillfully engineered a fiber optic based cable television system, which it called the Mini Hub. This system would not only provide cable TV, it could also let the subscriber have additional services such as security and fire monitoring. Times-Fiber was so committed to the technology that it actually set up a shop to manufacture its own optical fiber.

The problem was that fiber was not costing in for these types of applications, at least not in the early 1980s. For one thing, the subscriber did not receive the benefits of less repeater spacings that one got with the longer haul applications. For another, installing fiber in apartment complexes meant greater amounts of splicing and connectors—and connectors were not cheap, splicing could be imprecise, and these functions were better performed on coaxial cable. Nor did subscribers want to pay the extra prices for these so-called enhanced services. When the French Biarritz fiber-to-the-home project was completed, for example, only those living in 3 of every 10 homes connected paid for additional services such as videophone.

The result was that Times-Fiber had to scale down significantly its emphasis on fiber and offer a coaxial-based system as its primary network. "We had to go from a Cadillac to a Volkswagen," lamented then Vice President of Finance Kirk Evans.

That lesson was similar to the one learned at Hi-Ovis and Elie-Manitoba: Wiring localized environments with fiber optics to provide multiple services was extremely expensive. The result was that it was "now easier and often cheaper to communicate between cities than it is to communicate between floors," Trellis Corporation Vice President Allen Kasiewicz told the 1986 KMI conference.[4]

Yet the trend was to increase the use of fiber for shorter links as the price of fiber optic components decreased. By the end of 1986, the economic cutoff point had shrunk to 2 miles for customers with capacity requirements greater than 1.5 Mbps.

"The rewiring barriers to fall are universal understanding, standardization and fundamental economics," Kasciewicz said.

By the mid-1980s, both large corporations and small businesses recognized that the telecommunications industry represented a multibillion dollar opportunity, Bigham told the 1986 KMI conference. "This competition is helping to reduce the price tag of a fiber installation. Reduced prices aid in bringing fiber technology close to the central office."

Rewiring America's countryside was to cost out economically as the first large-scale application. By 1986, however, KMI President John Kessler was noting that "the future for fiber is in the subscriber loop portion of the telephone plant."

Early efforts to use fiber optics for localized data transfer also faced considerable cost barriers. One advantage was that the tremendous capacity that fiber afforded could do things such as unite two mainframe computers, allowing them to operate as one. Another advantage was that fiber's immunity from interference obviated office and factory noises plaguing twisted pair and copper-based networks.

On the other hand, systems designers were aware of the coming of the all-digital worldwide network—the Integrated Services Digital Network (ISDN)—and were attempting to plan that into their strategies. And these services required more than the standard channel bandwidth that traditional cables were geared to provide.

Computers are digitally based and fiber optics is basically a

---

[4]"Expanding Markets for Fiber Optic Systems in Buildings," by Allen Kasiewicz, presented at the 1986 KMI conference.

digital technology. The fact that fiber optics employs pulses of light using the binary code of ones and zeros makes it a perfect match for ISDN. This was not true of its competitor technologies—coaxial cable, satellite, and microwave—which had basically sent messages over analog signals, but which were upgrading their technologies to employ digital.

AT&T recognized this commitment to go digital and announced that it intended to have an all-digital network employing large amounts of optical fiber operating throughout the United States by 1990. When AT&T thought of fiber optics, it was becoming increasingly evident that it was thinking in terms of worldwide connectivity, rather than just rewiring the United States. The fact that fiber optics was gaining in worldwide acceptance was another factor to consider.

In the early 1980s, a consortium of 28 telecommunications representatives announced plans to build a trans-Atlantic fiber optic network operating by 1988 that would cost approximately $335 million. The network would link Tuckerton, New Jersey, with Widemouth, England, and Penmarch, France. The network is known as TAT-8, representing the eighth trans-Atlantic cable crossing, but the first using fiber optics.

The communications engineer now had four alternatives when it came to providing long-distance transmission, including fiber optics, coaxial cable, microwave, and satellite. The proponents of fiber optics felt that the technology could stack up equally well against each of these competitor technologies.

"Fiber is twenty times lighter, five times smaller, and, at 10 megahertz, has half the attenuation of copper lines of the same capacity," Schiller said. "It requires less signal regeneration than copper, is immune to electromagnetic interference, and is much easier to install. The high capacity and small size of fiber optic cable make it a natural candidate for high-density underground trunk routes through congested metropolitan areas."

"We have reached the point where the use of optical fiber is as economical as the use of coaxial cable, although variables still determine the better use," John Kessler said in 1983. "Factors include the cost of shielded copper wires and the distances they have to go," he continued. "The state of the technology today favors fiber because electromagnetic interference is increasing and the cost for metallic shielding is going up."

There are two services that simply cannot be transmitted via copper, according to Bigham. These are video and DS-3.

Fiber "stacks up equally well" against microwave transmission, said Schiller. "Traditionally, microwave has provided a cost-effective solution to long-haul large information capacity channels. Fiber optic transmission systems with repeaterless ranges up to 40 kilometers are ideal for this application."

Fiber optics was also beginning to take on the glamour technology of yesteryear—satellite. Then FCC Chairman Mark Fowler stunned an International Satellite Television Symposium in early 1983 by explaining that fiber optics should be the preferred medium for international television transmission. Fiber optics represents the next step in the march of transmission media, Fowler said, and the FCC must ensure "that this march of improved technology is not frustrated."

"Many countries have indicated their future networks will be based on fiber optics, and steps are already under way to achieve this end," Fowler told the satellite buffs. "Fiber optic cable may not prove to be the most advantageous transmission medium for all delivery routes," he said. But, "its technical characteristics of speed, increased capacity, no-time-delay variations, low susceptibility to interference or interception, and freedom from competition for orbital slots," make it a sound alternative.

The advantages fiber provided for long-distance routes were beginning to threaten satellite for domestic service in the United States. For example, the IBM subsidiary Satellite Business Systems was already beginning to talk about the advantages of fiber optics, a departure from the satellite technology on which the company was founded and from which it derived its name. One SBS official noted in 1984 that: "In the near future, the S for satellite in SBS will mean about as much as the T for telegraph means to AT&T."

In fairness, satellite offered the advantage of not having to buy or lease large sections of right of way, and laying cable was capital intensive. In fact, several carriers including MCI and Williams Telecommunications became involved in deals to sell their networks and then lease back the capacity to soften the financial blow. MCI did this on a number of routes. A basic advantage fiber optic carriers had was that they owned the capacity, rather than just leasing it from the satellite providers.

"For many applications," Schiller said, "there need be no other consideration than the fact that fiber optic based systems are cheaper to install and operate than traditional systems, and this trend will continue. Further, the rapid development in fiber optic cable production is helping to lower manufacturing costs."

Corning's L. C. Gunderson presented a paper at OFC '77 predicting that cost reductions to optical fiber "of more than an order of magnitude" should be possible. Gunderson's prediction became a reality only five years later, as the price of optical fiber had dropped from $3.00 per meter to about 30 cents per meter by 1982. "And not only has the price dropped," explained KMI President Kessler, "but also the performance of the fiber has improved significantly."

Gunderson told OFC '77 that "history has shown that, no matter how revolutionary or appealing a new technology may be, it is only incrementally economically advantageous at the time of widespread introduction."[5]

There was a feeling of inevitability about fiber optics, that its day was coming—whether that was going to happen tomorrow or in a decade, it was most assuredly going to happen.

The market for fiber optics in the United States in 1983 was pegged at $300 million, and this was to increase threefold, to nearly $1 billion, in the next three years. The United States represented the bulk of the worldwide market for much of the 1980s, accounting for two-thirds of that market in the early

[5]"Optical Waveguide Cost Considerations," by L. C. Gunderson, presented at OFC '77.

years, but predicted to command at least 50% in the years closing out the decade. This obviously represented an advantage to U.S. suppliers of fiber optic products, but the interest of foreign companies and governments eliminated any smugness U.S. suppliers may have derived from that fact.

"Foreign governments have targeted fiber optics as an industry they want their firms to dominate," explained a Department of Commerce analysis in 1984. "U.S. fiber optic manufacturers now face a situation where other nations have given this technology much more emphasis and visibility than in the U.S."

While noting that U.S. suppliers have "inherent advantages" in supplying to the U.S. market, the study says those manufacturers face "serious challenges" from foreign suppliers. The fact that other markets are not as open as those in the United States represents another problem for U.S. suppliers.

Marketers divided the basic cost of a fiber optic system into three parts: 1) transceivers (transmitters and receivers), 2) fiber/cable, and 3) connectors. Of those, fiber/cable represented the majority of the market. By 1983, approximately 455,000 km of optical fiber cable was being shipped annually in the United States, according to Kessler. This was to reach almost 2 million fiber km by 1986. (See Figure 1.)

An encouraging factor for the systems designer was that a plethora of suppliers had come about almost overnight. Western Electric had grabbed the early lead in manufacturing optical fiber, with Corning running a close second. A number of opportunities for other companies were also evolving—particularly in the area of manufacturing specialty fiber. By 1983, there were already 50 companies that could cable the optical fiber being manufactured, 12 of whom Kessler identified as being major. That market was diversifying as well, although again Western Electric was leading, followed by Siecor Corporation.

There were 16 U.S. suppliers of transceivers for fiber optics' use in telecommunications and 40 U.S. suppliers of transceivers for fiber optic data links. There were also 40 suppliers of fiber optic connectors to the U.S. market.

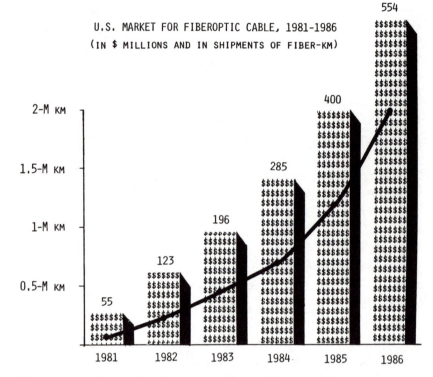

Figure 1. As the market for optical fiber cable grew, the amount used rose both in volume and market dollar value. This chart illustrates both trends. (Diagram courtesy Kessler Marketing Intelligence.)

New York Telephone was one of the Bell Operating Companies that was aggressively installing fiber optics. The telco had worked with AT&T in installing significant portions of the Northeast Corridor through its service area and had earlier helped with the 1980 olympics installation at Lake Placid. In 1982, New York Telephone installed its Ring around Manhattan, a fiber optic loop connecting 12 central offices in southern, central, and midtown Manhattan. (see Figure 2.) New York Telephone was wiring many of New York's banks and brokerage houses with fiber optics as part of the plan. This demonstrated that carriers were interested in fiber optics for shorter loop applications, as well as for the long haul.

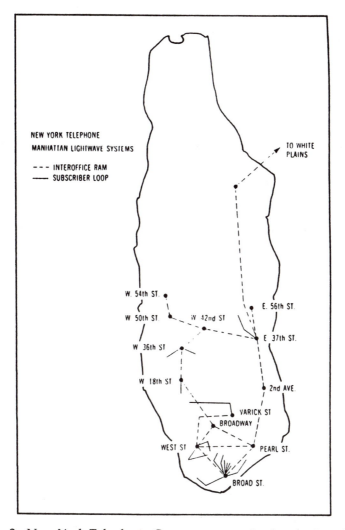

Figure 2. New York Telephone Company recognized early the advantages of using fiber optics. The telco joined central offices together with fiber in what became known as its "Ring around Manhattan." (Diagram courtesy New York Telephone).

"For the short loops typical of the urban environment, the fiber properties that prove advantageous are not long repeater spacings and high cross-section capacity, as in trunking applications," said New York Telephone's Frederick Frintrup. "Instead, the lightwave system characteristics to be emphasized include: small cable size, which avoids problems of congestion in underground ducts and building riser conduits; security of the transmitted information as compared to radio or copper wire media; immunity to electromagnetic noise interference; near-error free performance in well-planned systems; and protection switching over diverse routes that can only be efficiently provided over wideband systems."

If the Bell Operating Companies were finding out about the advantages of fiber optics throughout their areas, they were also going to be free to use whatever supplier they wanted because of the coming divestiture of AT&T on January 1, 1984. Divestiture was to further fuel the optical fiber furnaces starting to burn so brightly throughout the land.

Chapter 6

Divestiture's Impact

EVEN BEFORE ONE part of AT&T was helping to pioneer the creation of the laser and the resulting birth of fiber optics, a different department within AT&T found itself enmeshed in a separate struggle. This was the legal department, and it was locked in combat with a federal government intent on rending AT&T asunder.

The U.S. Department of Justice first brought suit against AT&T in 1949, claiming that AT&T was violating the Sherman Antitrust Act. The Department of Justice claimed that AT&T was conspiring with its subsidiary, Western Electric, to restrain trade in the manufacture and sale of telephone equipment. The suit requested that AT&T divest itself of Western Electric and that Western Electric give up its 50% ownership in Bell Laboratories.

The suit never came to court, and was settled in 1956. AT&T was allowed to keep its component parts together, but was told to restrict itself to "common carrier communications services." In return, AT&T agreed that it would not get into other lines of business, such as selling computers. AT&T also agreed to license all of its patents to all other companies, domestic and foreign. The result was that competing companies could come into the telephone market, but AT&T could not move out into other mar-

kets, a sort of "fence with a one way hole," as it was characterized at the time.

One example was the development of the transistor, which
AT&T had pioneered. A host of other companies—including a
number from Japan—were finding out about the miracle of the
transistor by applying AT&T's research; AT&T had its hands tied
behind its back and mainly had to stay out of the emerging new
product sectors.

Competitors began moving aggressively into the telephone
business in the spirit of the 1956 agreement and continued to
attack this market through the 1960s and into the 1970s. Antitrust suits continued to be filed against AT&T, including action
by growing competitor MCI. These led the Justice Department to
again move against AT&T in 1974, bringing suit against AT&T,
Western Electric, Bell Laboratories, and the 22 Bell Operating
Companies.

As in the first Justice Department suit, negotiations between
the parties stretched on for a considerable length of time. But
while an out-of-court settlement had been reached after seven
years in the first case, this time AT&T and the government were
still at loggerheads at the end of seven years. U.S. District Court
Judge Harold Greene brought the case to trial on January 15,
1981.

None of this had anything to do with AT&T and the growth
of fiber optics, at least within the AT&T infrastructure that was
actually developing the technology. The creation and incubation
periods went undisturbed, and AT&T had already begun using
its Bell Operating Companies as testing grounds for fiber optics.

Outside of AT&T, however, the general clamor for a more
open telecommunications marketplace was spurring competition. Suppliers other than AT&T were beginning to see the
potential for serving an enormous new marketplace—the Bell
Operating Companies, which provided local exchange service
throughout the United States and accounted for almost three-
fourths of the Bell System's total assets.

Justice Department Assistant Attorney General William Baxter told AT&T in April 1981 that the Justice Department intended to demand the divestiture of AT&T from its Bell Operating Companies. In return, the Justice Department was promising that it would release AT&T from the 1956 order that prohibited it from marketing other products, such as computers. Baxter was not pushing an oft-made threat of taking AT&T's manufacturing arm, Western Electric, away from it.

The threat of losing its Bell Operating Companies had an impact on AT&T and its development of fiber optics in the sense that AT&T did not necessarily have a huge captive market waiting for it anymore. While there were certainly synergy and relationships between AT&T and its Bell Operating Company subsidiaries that had been built up over decades, the Bell Operating Companies would also be free to look to other suppliers, should they want to.

Following Greene's rejection of a motion for dismissal by AT&T of the entire case in autumn 1981, the litigants got serious about striking a deal, and it ended up being the deal that Baxter had originally hoped to make. An arrangement was made as 1982 arrived. Greene concurred and, with modifications, issued the opinion in August 1982. The result was that AT&T agreed to divest itself of the 22 Bell Operating Companies, which were then formed into seven Regional Holding Companies, including: NYNEX, Bell Atlantic, Ameritech, Pacific Telesis, Southwestern Bell, BellSouth, and US West. The financial impact was somewhat devastating.

In 1983, AT&T's earnings were $1.8 billion on total assets of $150 billion. This placed AT&T in a position where it was larger than General Motors, IBM, General Electric, and U.S. Steel combined. Divestiture by AT&T of its Bell Operating Companies occurred on January 1, 1984. For 1984, AT&T reported earnings of $450 million on total assets of slightly less than $40 billion.

The breakup in effect created two competitive areas: the long-distance and local exchange markets. The country was divided into 161 local access and transport areas (LATAs). The seven Regional Holding Companies could provide service within the

LATAs in their prescribed areas; AT&T was now a long-distance carrier that provided service throughout the country to those areas. The two markets were joined by interconnections at the LATA boundaries. For the long-distance carrier to gain access to the local exchange area in which the Regional Holding Companies operated, the long-distance carrier had to pay an access charge. Since AT&T already had prime connections, it had to pay a greater fee at the beginning, but it had better quality than the other long-distance carriers. The goal of equal access thereafter was to provide all the long-distance carriers with the same service over the local exchange network.

AT&T used the opportunity to get into the computer and information services business by creating AT&T Information Systems (AT&T-IS), which was originally called American Bell. This organization, at least at the onset, was to maintain activities separate from those of the rest of AT&T, whose mandate it was to provide communications.

Bell Labs was impacted in two ways. First, 451 Bell Labs people were transferred on July 1, 1982 to form the Consumer Products Laboratories portion of American Bell. Six months later, 3559 employees were transferred to what had by then become AT&T Information Systems, with the term "Bell" taken from AT&T by the divestiture court with the exception of AT&T Bell Laboratories. Second, in order to provide R&D for the Regional Holding Companies, a separate company was spun out of Bell Labs. Originally known as the Central Services Organization, it was to become Bell Communications Research (BellCore). (See Table 1.)

Bell Labs was reduced in size from approximately 25,859 employees at year-end 1982 to 19,280 people at the end of 1985. Officials there insist their research and development activities in fiber optics have not been hampered by the changes. They maintain that the Labs' three primary research areas continue to be fiber optics, software, and microelectronics. The general consensus, according to an Annenberg study on divestiture funded in

Table 1. Bell Labs Employment History (year end totals). Bell Laboratories received a number of jolts from divestiture after having grown during the 1960s and 1970s. Included were personnel shifts both to other entities within AT&T and outside of the new AT&T. (Source: The Annenberg School of Communications).

	Year	Research	Total	Research as % of Total
Bell Labs	1965	1,011	14,517	6.96
	1966	1,054	14,402	7.32
	1967	1,089	15,229	7.15
	1968	1,115	15,938	7.00
	1969	1,066	16,442	6.48
	1970	1,057	16,998	6.22
	1971	1,222	16,768	7.29
	1972	1,239	17,187	7.21
	1973	1,322	17,064	7.75
	1974	1,306	16,491	7.92
	1975	1,311	15,989	8.20
	1976	1,303	16,137	8.07
	1977	1,328	16,932	7.84
	1978	1,375	18,229	7.54
	1979	1,405	19,454	7.22
	1980	1,424	21,955	6.50
	1981	1,227	24,078	5.10
	1982	1,249	25,859[a]	4.83
	1983	1,480	22,188[b]	6.67
AT&T Bell Labs	1984	1,163	19,333[c]	6.02
	1985	1,191	19,280[d]	6.18

[a] 451 employees transferred 7-1-82 to American Bell to form the Consumer Products Laboratories.
[b] 3,559 employees transferred 1-1-83 to AT&T Information Systems.
[c] 3,143 employees transferred 1-1-84 to Bell Communications Research.
[d] 1,173 employees transferred 5-1-85 to AT&T Information Systems.

part by the National Science Foundation, is that the Labs continues to be "reasonably healthy."[1]

[There is ample evidence to suggest that—regardless of the upheaval—AT&T has taken a very protective attitude toward fiber optics throughout the technology's history. When AT&T announced, for example, that it was going to lay off 27,400 employees in 1986, officials there made it clear that the layoffs were in no way going to affect the fiber optics program. The layoffs were announced, in fact, just as AT&T was putting the finishing touches on its nationwide fiber optic network and going forth with a bold new multibillion dollar program to install more cable.]

BellCore actually was formed by the transfer of 3400 AT&T people on January 1, 1984. [Included were both Bell Labs' and Western Electric employees.] By year-end 1985, BellCore had a total of 7713 employees. To some extent, BellCore is a competitor of Bell Labs. It has done extensive work in fiber optics, specifically concentrating on bringing fiber optics to the local exchange to aid the seven Regional Holding Companies. As part of its work, BellCore is attempting to improve photonic switching, with the ultimate goal of making switches and opto-electronic products mainly optical.

While the RHCs are mandated to pay for BellCore's services until 1989, they can begin dropping off at that point, which places pressure on BellCore to perform in the interim. One RHC—NYNEX—intends to create its own internal R&D organization of about 350 people by 1990. For its part, Pacific Bell started up a group to track and assess technology. US West has let it be known that it may not reinvest in BellCore when the contract lapses in 1989, and other RHCs have at least toyed with the idea.

---

[1]"The impact of divestiture on Bell System Research and Development Activities," by A. Michael Noll, The Annenberg School of Communications, University of Southern California, 1986.

While the RHCs were not initially allowed to manufacture their own products, they are trying to gain entry into that market, and are straddling the line in various instances. For example, BellSouth has purchased a 50% share in FiberLAN, which formerly was a wholly owned subsidiary of Siecor Corporation. FiberLAN's stated goal is to develop fiber optic based local area networks. It had been a supplier to various companies in the past. Officials claim that FiberLAN does not itself manufacture products, but rather designs systems and puts components together.

One potential bend in the river as post-divestiture activities continue to flow forward is that the RHCs may yet be allowed to manufacture products and enter the interexchange markets. A report issued by the Justice Department in early 1987 has recommended just that. One strong argument in favor of allowing them into the manufacturing arena is that it may help to curtail the invasion of foreign-manufactured products.[2]

Allowing the RHCs to enter the interexchange carrier market would add a new group of potential carriers to compete in an already crowded marketplace. The idea has drawn the wrath of various long-distance fiber optic carriers. The National Telecommunications Network, for example, proposed a six-year ban on allowing the RHCs into the long-distance business in early 1987.

For all the trauma and second-guessing that divestiture has wrought, the operating units that made fiber optics a reality at AT&T still remained intact after divestiture. Bell Labs is still in existence, although in somewhat altered form, and it works directly with other AT&T operating units, including personnel from what was formerly Western Electric. (See Figure 1.) This is no doubt a relief to the military, which is very dependent on

---

[2]"Report and Recommendations of the United States Concerning the Line of Business Restrictions Imposed on the Bell Operating Companies by the Modified Final Judgment," (Civil Action No. 82-0192), released February 2, 1987.

Figure 1. The strong working relationship between Bell Laboratories and Western Electric continued mainly unchanged through divestiture. This predivestiture photo shows John Janowski checking optical fiber cable used to transmit telephone traffic, control messages, and timing information within the No. 5ESS AT&T switch. A Western Electric engineer, Janowski was on loan to Bell Laboratories at the time. (Photo courtesy AT&T Bell Laboratories.)

AT&T for transmitting classified and otherwise important military communications. The infrastructure between AT&T and the Bell Operating Companies that existed prior to divestiture in many ways has stayed the same in this regard as well.

Despite these continuing associations, one must wonder whether fiber optics as a technology will continue to move forward in the post-divestiture era with a clipped Bell Labs. One impact of divestiture on Bell Labs has been that Bell Labs' role "as a national telecommunications R&D resource will probably diminish," according to the Annenberg study. Bell Labs certainly acted as a national resource—even a world resource—as it created and developed fiber optics.

Whatever national technological direction Bell Labs provided over the decades must now be shared with BellCore. "With divestiture, Bell Labs to a considerable extent lost its mission to assure the future of telecommunications in the United States and also lost the stability of funding that came from a business that was predominantly providing a service rather than selling products," says the Annenberg study. "To some extent, Bellcore has inherited that mission along with a funding base that is based on the provision of service. Thus, Bellcore might emerge as a new national resource in communications-related research and development."

Yet another concern is whether the diversification of the R&D effort—now including Bell Labs, BellCore, and RHC groups—might lead "to fragmentation and a loss of synergy across research disciplines," according to the Annenberg study.

AT&T has retained most of its captive market with the Regional Holding Companies in the years immediately following January 1, 1984. It is difficult to pry away a grip that has developed over so many years. For example, AT&T still works extensively with Bell Operating Companies in experimenting with ISDN. Some have observed that ISDN may serve to negate the impact of divestiture by bringing AT&T and the Bell Operating Companies together again, since together they have the infrastructure and revenues to fund something as intensive as ISDN. In fairness, the RHCs are also dealing with other manufacturers on ISDN.

With regard to providing communications services, however, AT&T not only lost its local exchange carriers, but also faces increasing competition in the long-distance marketplace. Companies such as GTE Corporation, United Telecommunications, and MCI have as much right to tie up with the local exchange carriers as does AT&T.

This has hastened the rewiring of America with fiber optics. The so-called other common carriers must build fiber optic systems and "install optical fiber in much greater amounts than the actual need," KMI's Kessler explained.[3] The reason quite simply is that fiber optics—with capacity a carrier can grow into—affords these organizations with the only means to compete successfully against AT&T in the long-haul market.

AT&T in the years after divestiture complained that it had to arduously file with the FCC each time it filed an intention to build or upgrade a fiber optic link, while a filing by one of the other long-distance carriers was a matter of little FCC interest. Of much greater importance, AT&T's rates remained regulated. All this came at a time when AT&T was trying to change its image, attempting to become leaner and meaner, so that it could compete effectively both as a carrier and as a supplier.

However, the other common carriers reminded the FCC that if anything was done to topple the apple cart and deregulate AT&T too rapidly, the other carriers would be put out of business. The various other carriers had to raise extensive amounts of capital for building fiber optic routes, for example, and had to keep AT&T regulated until those routes were to begin generating significant business.

The other competitive market established by divestiture was in the local exchange. While the Regional Holding Companies controlled approximately 90% of the traffic in their respective areas prior to divestiture, following divestiture they were in a position where they had to fight hard to keep those markets.

While they reaped the benefits of the lucrative access charges the long-distance carriers had to pay them, the RHCs also ran the risk of forfeiting some of that business if the long-distance carriers found ways to more economically connect with alternative carriers. When a company uses private or leased transmis-

---

[3]"Exclusive Interview: KMI's Kessler Predicts $3 Billion Market by 1989," *Fiber/Laser News*, July 22, 1983, Phillips Publishing Inc., Potomac, Maryland

sion facilities to avoid the local telephone company network, that group is said to engage in bypass.

The RHCs are speeding up their involvement with fiber optics in order to slow down or eliminate altogether the threat of bypass. In some instances, bypass carriers are offering fiber optics capacity, putting the heat on even more. This generally has spurred the growth of fiber optics in the local exchange area.

For 1984, the first year following divestiture, Siecor Corporation estimated the Regional Holding Companies had installed a total of 250,000 fiber kilometers, led by BellSouth (55,000 fiber km), NYNEX (40,000 fiber km), and Ameritech (40,000 fiber km).

One Ameritech subsidiary, Ohio Bell, announced in 1984 that it was going to spend $87 million over a three-year period to install fiber optic transmission equipment. That was because "we're not going to wait for competitors to take business away from us," explained Vice President for Network Services Donald Baker.

In addition to NYNEX's "Ring around Manhattan," other RHCs were also realizing the importance of bringing fiber optics into major metropolitan areas. A separate Ameritech subsidiary, Illinois Bell, built and began operating Nova-Link, which incorporated large amounts of optical fiber cable in the Chicago area.

The fact that AT&T no longer controlled the rate at which fiber optics was incorporated into the networks may also have had a liberating effect on the RHCs, which certainly have qualified engineers of their own knowledgeable about fiber optics.

Despite this aggressive attitude by the RHCs toward fiber, the growth of bypass had begun and its impact is "terrifying," explained Martin McDermott, vice president and general manager of the National Telecommunications Network (NTN). Bypass, says McDermott, will be "pretty ecumenical," crossing throughout the territories of various Regional Holding Companies.

NTN's own operating philosophy was to first go to the RHCs

and request local exchange access within 60 days. If something could not be worked out in that time, NTN would begin talking to alternative carriers.

The bypass carriers come from a variety of areas. Cable television oriented companies such as Warner Amex have been aggressive in their involvment. Other companies, such as Dama Telecommunications, have devoted themselves to providing bypass as a major part of their corporate undertaking.

The rejuvenated Regional Holding Companies have also longingly looked at the cable television market as an area in which they would like to get involved. While restricted from providing enhanced services of that kind, the RHCs have been able to come in as contractors to cable television franchises to build a network and have already done so, including fiber optic routes.

One burgeoning customer base for bypass services is the Fortune 500 companies, which are interested in controlling their own communications facilities and obviating the need for large phone bills that otherwise would be paid to the RHCs. The bulk capacities these companies often require in specified geographical locations make tailoring their networks almost an economic fact of life.

Despite NYNEX's aggressive efforts, successful bypass carriers began cropping up in the New York City area. Because it carries the largest amount of traffic in the United States, New York City has suggested itself as a good early market for bypass companies if they have the operating expertise, the right-of-way, and the means to deal with the disparate groups that make the city run.

In early 1984, AT&T warned the New York Public Service Commission that AT&T might be forced to bypass New York Telephone's local network if long-distance carriers were charged exorbitant access charge fees. In response, the Public Service Commission pooh-poohed the charges, with one official unwittingly noting that bypass "would take a lot of planning and logistics." New York Telephone President William Ferguson, however, acknowledged the validity of AT&T's claims, explain-

ing that pricing and regulation actions were leading to "artifi-
cially tough prices that could lead to uneconomic bypass."

Soon thereafter, AT&T announced that it had reached an
agreement with Teleport Communications, Inc., for Teleport to
provide local service to AT&T in Manhattan. In effect, this meant
that AT&T would be bypassing New York Telephone. Teleport
Communications is 97% owned by Merrill-Lynch, Inc., and 3%
owned by Western Union. Teleport seems destined to become a
successful large-scale bypass provider, operating a 150-mile
regional fiber optic network throughout New York and New Jersey, and tying into earth stations on Staten Island. (See Figure
2.)

That AT&T had signed an agreement with a bypass carrier
meant to some that there was growing friction between AT&T

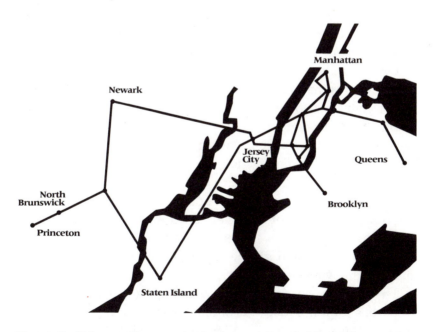

Figure 2. Teleport Communications was not afraid to take on New
York Telephone as an alternative carrier. The map shows the 150-mile
service area covered by Teleport's regional network. (Map courtesy
Teleport Communications.)

and some of its former subsidiaries and to others that divestiture was beginning to work as it should. The Regional Holding Companies, for their part, were certainly not obligated to use only AT&T and were in fact mandated to provide equal access to those who wanted it.

A separate bypass provider in New York, Cable & Wireless subsidiary TDX Systems, linked four Manhattan locales with a 52-mile fiber optic bypass.

As we will explore in Chapter 8, the bypass providers who had found out about fiber optics were using it as a selling point to compete against the Regional Holding Companies. They claimed, for example, that their networks were all-digital. The RHCs, of course, were to some extent stuck with their existing networks, which were mainly analog based.

There was also resentment aimed at the RHCs for not trying to concentrate all of their efforts on upgrading their own networks rather than trying to diversify into other areas, which some of them were doing.

Divestiture was in many respects a unique experiment in the United States that has become a catalyst for change and growth. Other countries, following that example, began doing the same thing to a more limited degree, including Japan and Great Britain. In both instances, fiber optics is growing in use and the marketplaces have been enlivened.

Not everyone would agree that divestiture has had a positive effect, particularly many residential users, who have suffered discord that would not have happened when AT&T was whole. Divestiture has, however, served as a springboard to further accelerate the growth of fiber optics in the most open marketplace in the world—the United States.

The commitment as of 1984 was being made to fiber by a host of parties for a number of economic reasons; it had spread well beyond the early grasp of AT&T and Corning. The conditions were right for installing large amounts of fiber optics—the time had come for the rewiring of America to take off.

Chapter 7

Rewiring the Countryside

A WHOLE NEW world had opened up in the United States to tele-communications providers, and it was clear that fiber optics was playing an essential role in their strategies.

While AT&T and MCI had begun 1983 in heated competition, it soon became apparent that others were coming into the long-distance fray, and they were using fiber optics as their main weapon. But to compete effectively, carriers needed the right-of-way to install fiber optics—here they found a new friend that had been around for a long time.

MCI had already realized the importance of doing business with the railroads. It was Amtrak—by leasing its right-of-way—that allowed MCI to connect Washington, D.C., with New York City by fiber optics. CSX allowed MCI the prospect of installing fiber optics on up to 4000 miles of right-of-way for 25 years at a cost of $32 million, providing a gateway to MCI throughout the eastern United States.

The railroads, in response, were beginning to see the advantages of striking agreements with the fiber optic communications carriers as well. In a confidential report to the American Associ-

Figure 1. Much of the rewiring of America occurred alongside railroad tracks. The railroads represented a tremendous source of right-of-way for carriers seeking to connect cities together with optical fiber cable. For the railroads, the rewiring represented a new source of revenue. (Photo courtesy US Sprint.)

ation of Railroads, the railroads were urged to team up with the common carriers for mutual benefit.[1] (See Figure 1.)

For one thing, providing right-of-way to these carriers meant money to the railroads, many of whom were ailing financially. And the carriers were prepared to spend large amounts in the pursuit of fiber optic systems. MCI, for example, had apportioned more than $150 million in its fiscal 1984 budget for fiber optics and estimated that it would cost $500 million to build its network throughout the eastern United States. (See Figure 2.)

The report to the AAR noted that the railroads would benefit from fiber optic systems by allowing the railroads to replace their

[1]"Confidential AAR Report Recommends Railroads Provide Rights of Way for Fiber Optic Use," *Fiber/Laser News*, June 24, 1983.

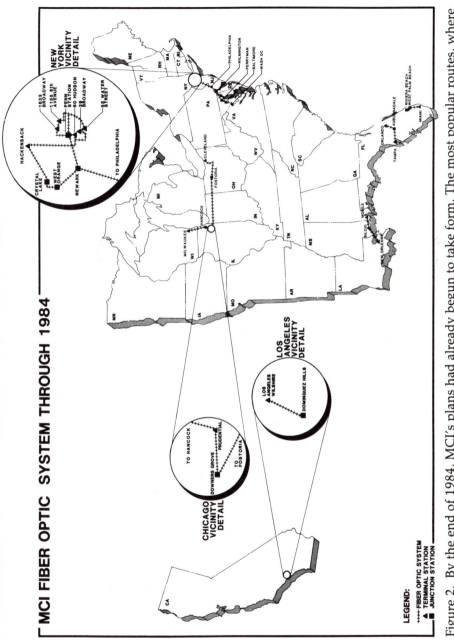

Figure 2. By the end of 1984, MCI's plans had already begun to take form. The most popular routes, where right-of-way was available, were serviced first. (Courtesy MCI.)

own microwave or leased circuits with fiber optic capacity. Fiber optics was destined to "improve dispatching, signaling, and computer controls."

There was also the irony that carriers were using an industry that had been around longer than AT&T had existed to allow them to compete effectively against AT&T. It spoke well of a capitalistic society that a budding new information industry could bolster a services industry on the decline.

While noting that railroad dealings with MCI had been for leasing right-of-way only, the report said that other carriers might be more "open-minded" and strike partnership deals with the railroads.

Other types of agreements railroads might want to enter into with fiber optic carriers included: presale and installation of optical fiber on "a condominium basis;" wholesale or "carrier's carrier" agreement, where capacity could be sold before capital costs were incurred; or as a retail carrier.

The study was to become an influential piece of work. The railroads took seriously the AAR study's commandment to gain a larger piece of the action than just leasing their right-of-way. For the next announcement to rattle the marketplace again involved CSX Railroad, but this time CSX had found a carrier that would act as a full-fledged partner.

It was from the top of the World Trade Center tower in New York City in the summer of 1983 that CSX, together with Southern New England Telephone Company, pronounced the birth of Lightnet. Lightnet billed itself not as a carrier operating like AT&T or MCI, but as a "carrier's carrier" allowing carriers to buy or lease the capacity along its fiber optic network. The venture was not interested in making its living from residential voice traffic; it was going to make a buck by providing carriers and large businesses with a way to expand their networks.

CSX and SNET were equal partners in the venture, a major departure from the lease arrangement MCI and the railroad had established. SNET was providing the capital and engineering expertise; the railroad was providing the right-of-way. As an article in *Business Week*[2] explained: the railroads themselves were not paying a penny to get involved.

The Lightnet network was originally announced as running 5000 miles through 22 states. Officials steadfastly maintained that no leg of the system would be built until all of the customers had been brought on board for that particular link.

Lightnet was ostensibly built to provide competition to AT&T and other carriers in supplying business services to customers in the long-distance market, although the carriers could also become Lightnet customers.

Yet AT&T was emerging with another hat on in the long-distance marketplace—that of supplier to other ventures—and had the inside track in providing both electronics and cable to Lightnet. As the announcement was made, it became apparent that Lightnet would have the same 90 Mbps electronics with potential upgrade to 432 Mbps that AT&T employed along its fiber optic routes.

It must be noted that Southern New England Telephone Company was 24% owned by AT&T at the time of the announcement and was considered an AT&T affiliate in a relationship that dated back almost 100 years. AT&T was to give up that share— and its affiliate role—not long after.

To further cement the relationship, Western Electric was named to provide the supplies for Lightnet's first leg, a 461-mile route stretching through the state of Florida. That route joined Miami, Orlando, and Jacksonville, with a spur to Tampa. AT&T also received additional contracts from Lightnet.

It is interesting to note here that—as fiber was coming to market from a variety of corners—relationships began to blur. AT&T was building extensive amounts of fiber optics along its own network, yet was acting as a supplier to Lightnet. CSX was in active partnership with Southern New England Telephone Company,

---

[2]"A Ready-Made Track for High-tech Communications," Business Week, September 12, 1983.

but was also leasing right-of-way to MCI to the same geographical locations.

The next major announcement, shortly after Lightnet's, was of a new venture to be known as Electra Communications. This joint venture again involved a major railroad, the Missouri-Kansas-Texas, which was linking up with the British based Cable & Wireless. Like Lightnet, Electra was regional in scope and was geared to the business and carriers community and not the residential market. Electra announced that the first leg of its network was going to service the dense route between Dallas and Houston, stretching 560 miles.

An important marketing distinction between Electra and Lightnet was that Electra intended to start building its network even if all the customers had not been lined up. That may have sprung in part from the commitment of Cable & Wireless chairman Sir Eric Sharp, who intoned from across the waters that hundreds of millions of Cable & Wireless dollars were earmarked for C&W to gain entry into the U.S. fiber optic marketplace.

As though these two networks were not enough, a third fiber optics venture was also in the works at about the same time. This was Microtel, which was to cost $60 million and traverse 800 miles of right-of-way provided by the Florida East Coast Railroad. The railroad would provide the right-of-way under a lease arrangement similar to the one MCI had with CSX.

Two supertrunks were planned for Microtel, one along each of Florida's two coasts. The east coast route joined Miami to Jacksonville, and included Ft. Lauderdale and Titusville. The west coast network connected Naples in the south with Lake City in the north, and was to join Ocala, Orlando, Tampa, St. Petersburg, Sarasota, and Ft. Myers. An east–west route joined Orlando and Titusville. Similar to Lightnet and Electra, but unlike AT&T and MCI, this was intended to be a regional venture. However, unlike Lightnet and Electra, Microtel was interested in providing residential telephone service. Microtel hoped to capture both discount telephone services and business oriented customers. It also promised the enhanced services of high-speed data, facsimile, teleconferencing, and videoconferencing.

"The union of Microtel and the Florida East Coast Railroad marks a truly significant step for the development of Florida's most sophisticated telecommunications system," announced Microtel chairman Richard Smith. Microtel trumpeted the fact that it was the first company in the United States to be granted a state certificate of authority to construct and operate an intra-state long-distance telephone service.

While it announced that it was beginning with the same 90 Mbps electronics that AT&T used, Microtel was very interested in providing state-of-the-art equipment.

While AT&T said its 432 Mbps electronics would be coming soon, the Japanese suppliers were explaining that they could provide the higher speed electronics sooner. While fiber optics was a brand new technology, Microtel—like MCI before it—realized that it had a choice. And it realized the importance of having the best equipment—not equipment that might be out-dated by the time it went into the ground.

The Japanese were coming to market and offering higher speed electronics much faster than AT&T. Consequently, Micro-tel went to Fujitsu and arch Fujitsu competitor NEC Corporation to try to get the state-of-the-art equipment that it desired. After a heated battle, Microtel went with NEC's 400 Mbps electronics, bringing a second major Japanese manufacturer into the commercial U.S. fiber optics marketplace.

What must have been frustrating for AT&T was that it continued to lead the world in technology—if one only considered hero experiments. In early 1983, Bell Labs had sent a repeaterless signal 119 km at 420 Mbps using single-mode fiber. Later that year, AT&T sent a repeaterless signal 161 km at the same transmission rate. The second experiment bested by 27 km a 134-km repeaterless signal transmitted by Nippon Telegraph and Telephone months before.

While AT&T had first brought fiber to the Northeast, and interest in that area continued with MCI's New York–Washington route, the venue was shifting rapidly to Florida. MCI was doing survey work along CSX right-of-way in Florida, with two of its first three routes destined to be built there. Microtel was building in Florida, and Lightnet had chosen to construct its first route there. (See Figure 3.)

Florida also became the entry point for another foreign supplier hoping to garner a piece of the U.S. fiber optics market. Chased out of the lucrative supplier position it had with British Telecom as part of BT's deregulation, Plessey, Ltd., both literally and figuratively, had to search for new worlds to conquer.

No doubt the most alluring market was the United States, and

Figure 3. An early focal point in the rewiring of America with optical fiber was Florida. In order to bury optical fiber cable deep enough, in effect keeping it out of the way of other equipment that may unknowingly damage it, it was necessary to first trench the ground where the optical fiber cable was going to be laid, sometimes to distances of 4 feet. (Photo courtesy US sprint.)

Plessey got its foothold by buying the Florida based Stromberg-Carlson's digital switching operations. Long respected as a digital switching supplier for telephony applications, Stromberg-Carlson gave Plessey an added legitimacy as it eyed the U.S. community.

The Japanese players were also establishing beachheads in the United States. Fujitsu announced that it was going to build a plant to manufacture fiber optic components in the Dallas area, as one Fujitsu official explained, "right around the corner" from where MCI had several facilities. [MCI was to later move its fiber optics transmission sector to the area.] NEC soon thereafter announced that it was going to build a facility in Virginia to manufacture fiber optics equipment.

Sumitomo Electric later in 1983 said it was going to build a major facility in Research Triangle Park, North Carolina, which would manufacture optical fiber.

Both Corning and AT&T had experimented with single-mode fiber from the beginning, and both had recognized its advantages throughout the period that fiber optics was being developed.

However, it was not until autumn 1983 that the first commercial single-mode fiber system was cut over. This was accomplished by a large independent telephone company, Continental Telephone of New York. It was a 37-km repeaterless link joining Norwich and Sydney, New York. ITT Telecom provided the 90 Mbps electronics while Siecor supplied the single-mode fiber cable. The link operated at the 1300 nm wavelength.

Days thereafter, MCI announced the second commercial single-mode optical fiber link, a 15.5-mile route joining Los Angeles and Dominguiz Hills, California. Perhaps even more important, however, was the fact that MCI was using the first commercialized high-speed fiber optic equipment, Fujitsu's 405 Mbps electronics.

The deal in some respects represented a dark day for U.S. fiber optic electronics suppliers generally and AT&T specifically. A concerted research effort by the Japanese had brought high-speed electronics to the U.S. market well ahead of the U.S. companies. Consequently, the U.S. market—at least temporarily—was surrendered to NEC and Fujitsu. AT&T's only rejoinder by the end of 1983 was to announce that it could increase its transmission rates to 180 Mbps, still not close to the 400 Mbps levels the Japanese were now delivering. The lesson for AT&T appeared to be that you can have the best operating transmission system in the world in your own laboratory, but if it's too delicate to operate in the field, what good is it?

The scenario was not unique to fiber optics, it had been played out in other technologies as well—the U.S. develops the R&D and the Japanese exploit the applications.

But while NEC and Fujitsu were successfully positioning themselves in the U.S. market for fiber optic electronics, Corning was doing its best to make sure the Japanese and other outside suppliers did not make inroads into the U.S. market for optical fiber.

Companies other than Corning or AT&T trying to sell competitive optical fiber were getting slapped with patent infringement suits by Corning, and Corning was winning. ITT had found out the hard way that it had to receive a license through Corning. Corning even sued the U.S. government and collected $650,000 because the government had bought optical fiber cable from companies such as ITT and Valtec Corporation without paying Corning royalties on the sales.

New fiber optic suppliers in the United States such as American Fiber Optics Corporation and SpecTran Corporation learned the lesson: pay Corning up front and get the matter settled early.

Despite Sumitomo's intention to build a plant in North Carolina, the threat of Corning legal action hanging over it significantly limited its ability to compete in the United States. Perhaps spurred by its Japanese counterparts, Sumitomo began advertising its product in the U.S. market, but its share was being limited to less than 5%.

Simplifying Corning's continuing patent battles was the later announcement that ITT had purchased Valtec from N. V. Philips. The Valtec venture had been plagued with instability; joint venture partner M/A-Com had bailed out at what sources said was a considerable loss, leaving the venture with N. V. Philips as sole owner. Philips, in turn, had searched for a buyer until ITT agreed to terms. Corning had sued Valtec for patent infringement; since Valtec was to be an ITT subsidiary, it would also fall under the Corning license ITT had negotiated. (Valtec was to be purchased again, this time by (CGE; Corning continued to maintain the patent rights.)

As 1983 came to a close, the rewiring of America's countryside was beginning in earnest. While the initial hotspots were in the Northeast and Florida, there were indications that the entire eastern United States was going to be wired. (See Figure 4.)

Figure 4. Early efforts to install optical fiber cable were performed mainly in the eastern United States. (Photo courtesy US Sprint.)

The technological achievements first hinted at in the laboratory almost two decades before were coming about. Single-mode fiber was beginning to make an impact, and its presence was to become overwhelming. High-speed electronics had also arrived, and by November Plessey said it was going to raise the transmission rates one notch higher by providing 565 Mbps electronics to British Telecom in the near future.

As 1984 rolled in, it became apparent that those companies already committed to building fiber optics throughout the land were accelerating their plans. It also became evident that new carriers were coming into the game.

One early entrant, MCI, spent nearly $1 billion on fiber optics in the period from 1983 through 1986, according to MCI spokesperson John Houser.

In March 1984, a midwestern group filed with the FCC to build a 1190-mile fiber optic route, including Chicago, Detroit, Toledo, Cleveland, Indianapolis, Cincinnati, Louisville, Lexington, Akron and Pittsburgh. The president of Litel, Inc., Lawrence McLernon, identified its mission as serving the "Fortune 1000 companies" located along that midwestern route. McLernon noted that the people working on the venture would be "career telephone people."

Soon thereafter, it was to become apparent that Litel was employing a strategy all its own in bringing in suppliers and financial partners. Pirelli Cable Corporation came in as an optical fiber cable supplier and equity partner. The Italian based Fiat subsidiary Telettra was to come in later, also as a supplier/owner. Negotiations with the French Societe Anonyme de Telecommunications also involved ownership, although SAT was to come in eventually as a supplier only.

United Telecommunications soon thereafter announced its intentions to build a nationwide fiber optic network. The company was prepared to spend $4 billion and build for 10 years in order to get the network completed. United was talking with every major railroad in the United States about providing right-of-way and holding discussions with just about every fiber optics supplier. (See Figure 5.)

United Telecommunications' first deal involved the Illinois Central Gulf Railroad, which was providing right-of-way for the first portion of this network, which was to run between Kansas City and Chicago. (See Figure 6.) The decision to build the nationwide network came following exhaustive research. United Telecommunications anticipated that voice traffic throughout the United States would increase fourfold between 1984 and 1994— from 150 billion call minutes to 600 billion call minutes. The network was to be single-mode fiber based and operate at high transmission rates.

The announcement brought to three the number of carriers considering the construction of nationwide fiber optic systems—

Figure 5. United Telecommunications announced that it was going to build a nationwide fiber optic network, even if the cost was going to reach billions of dollars. Here, a US Sprint link is being put in using General Cable optical fiber cable. (Photo courtesy US Sprint.)

Figure 6. United Telecommunications relied heavily on railroad right of way to help it carry out its ambitious plans to build a nationwide fiber optic network. Its first deal to gain right of way came with the Illinois Central Gulf Railroad. (Photo courtesy US Sprint.)

United Telecommunications, AT&T, and MCI. GTE was also installing cable to service areas it considered important. There were also four regional carriers, including Lightnet, Electra, Microtel, and Litel.

In spring 1984, Lightnet acknowledged that it had erred in developing a marketing strategy that dictated that it would not build a route unless all of its customers were signed up. Why should customers commit themselves to something that wasn't even built, particularly when they could obtain capacity along an existing route or a fiber optic route that was in the process of being constructed?

Following a significant layoff, which slowed the project, Lightnet began building. Its first link was the Florida network.

Still another regional fiber optic network was announced in June 1984. This was SouthernTel, whose initial plans called for construction of a 538-mile fiber optic network in North Carolina. The backbone system was to run from Charlotte to Rocky Mount, with other cities to include Asheville, Hickory, Winston-Salem, Greensboro, Durham, Chapel Hill, High Point, Raleigh, Sanford, Fayetteville, and Wilmington.

Meanwhile, AT&T continued to sprinkle fiber optic routes throughout the United States. It filed in June 1984 for a fiber optic network between Orlando, Florida, and Monticello, Georgia, running through Macon, Georgia. The system was estimated to cost $63.65 million, approximately $25 million less than if coaxial cable had been used, according to AT&T.

It soon became apparent that AT&T also was not adverse to using right-of-way provided by the railroads. The first such indication came when Union Pacific System said it was supplying less than 100 miles to AT&T in Texas.

The trade newsletter "Fiber Optics News" discovered that AT&T was using railroad right-of-way much more than it was letting on. The newsletter reported in its July 15, 1985 issue that AT&T was leasing right-of-way from Conrail, Florida East Coast Railroad, Union Pacific Railroad, CSX Corporation, Boston & Maine Railroad, Amtrak, and Grand Trunk Railroad. AT&T officials estimated that approximately 10% of AT&T's proposed fiber optic right-of-way would be supplied by the railroads.[3]

AT&T also announced it was going to use Conrail and New Jersey transit to provide right-of-way for a single-mode fiber optic network between Philadelphia and New York City. AT&T claimed that "excessive lease charges" by Bell Atlantic forced it to go with the rails. The single-mode fiber route was to run parallel with the Northeast Corridor installed earlier and

---

[3]"AT&T is revealing strong dependence on outside sources to provide fiber right-of-way," *Fiber Optics News*, July 15, 1985.

was needed to provide capacity once the multi-mode fiber was filled.

Union Pacific also announced that it was a major supplier to United Telecommunications, providing access to 5400 miles across 14 states, including: Arkansas, California, Colorado, Idaho, Kansas, Louisiana, Missouri, Nebraska, Nevada, Oregon, Tennessee, Texas, Utah, and Wyoming.

MCI also announced that it had worked out a deal with Union Pacific to gain access to 2300 miles of right-of-way traversing Missouri, Kansas, Arkansas, Texas, Louisiana, California, and Colorado. This was in addition to 375 miles of right-of-way that MCI had also negotiated to lease from Illinois Central Gulf Railroad.

MCI had also come to terms with the Boston & Maine Railroad to gain access to 300 miles of right-of-way and the Chicago-Milwaukee-St. Paul and Pacific Railroad for 78 miles of right-of-way.

Still another regional fiber optic venture was born in what was becoming a phenomenon of majestic proportions. This was LDX Net, a St. Louis telecommunications company whose major owner, Kansas City Southern Industries, also owned the Kansas City Southern Railroad. The venture aimed to build a fiber optic network throughout the Midwest and Southwest, and was also using right-of-way provided by Union Pacific.

LDX Net hired AT&T and Anaconda-Ericsson to provide its optical fiber cable. A prime mover in the venture, Don Hutchins, explained that the numerous fiber optic ventures sprouting up were creating a shortage of optical fiber cable, which made it essential for one to order optical fiber cable where one could, not necessarily where one wanted to. (See Figure 7.)

Realizing demand was overshadowing supply, Corning announced that it intended to spend $87 million (which eventually came closer to $100 million) to expand its optical fiber cable manufacturing facility in Wilmington, North Carolina. Corning was gearing up to have the capability of manufacturing up to 2.3 million fiber km annually by 1986.

Figure 7. Gordon (Don) Hutchins, Jr., who was to become the carrier's president, was an essential element in the birth and rise of LDX Net. A member of the National Telecommunications Network, LDX Net nevertheless was to sell capacity aggressively on a nationwide business using its own people. (Photo courtesy LDX Net.)

As though the LDX announcement wasn't enough, word of yet another new venture—Fibertrak—began leaking out in the summer of 1984. This was to be a cooperative venture including the Sante Fe, Southern Pacific, and Norfolk-Southern Railroads, and its announced breadth and scope were something to behold. The venture was talking about building a nationwide fiber optic network that would use almost one billion meters of optical fiber, a dizzying sum even for these times. Questions were being raised about Fibertrak even as the announcement was being made, however. One central concern was where all the customers were going to come from to use the capacity that Fibertrak was proposing.

Filling up capacity became a driving force of these various regional and national ventures to get their systems up and operating as rapidly as possible. There was the fear that—especially if two carriers were aiming to serve the same geographical region—the first one to complete its route was going to be declared the victor. The loser could potentially be driven from the marketplace, the victim of the larger capacity that fiber optics provides, an advantage that it also had been trying to sell to customers.

MCI had cut over its New York–Washington fiber optic route earlier in the year. Approximately one year after its announcement, Lightnet said it had completed the first leg of its Florida route, the Miami–Tampa–Jacksonville section.

Further upgrades and announcements continued, almost on a monthly basis. SouthernTel merged with the Georgia based Interstate Communications, Inc. to form SouthernNet. SouthernNet's proposed network was to now include Virginia, North Carolina, and South Carolina. Former SouthernTel head Gene Gabbard was named SouthernNet president, and former ICI president Cam Lanier was named chief executive officer. (SouthernNet was to merge again, this time with TelMan.)

SouthernNet was joining hands with Microtel to form the Southeastern Telecommunications Network. The Norfolk-Southern Railroad had recently found yet another way in which railroads could interact with these fiber optic carriers; it now owned 20% of Microtel. Also included in these discussions was Rochester Telephone subsidiary RCI Corp.

But while RCI's role in these discussions was to provide capacity between Washington, D.C., up through the Northeast, the carrier had plans of its own to build a 580-mile fiber optic network to link Buffalo with Cleveland, Toledo, Detroit, and Chicago. Conrail provided the right-of-way for the RCI route. To interconnect its Buffalo network to New York City, RCI was going to build a large microwave facility.

The RCI Buffalo–Chicago fiber route was directly competitive with Litel's. RCI was in the secure position of being funded by a larger telephone company. Litel, however, was still wheeling and dealing to make the venture go. Other telephone companies were brought in to help fund it; Alltel now owned 32% and Centel owned 28%. Pirelli cable was in for 20% of the venture. But Litel was aggressively installing cable and had already announced its first customer, Firestone Tire & Rubber Company.

Litel had some success in going around the railroads to garner its right-of-way. It signed an agreement with the Ohio Turnpike through which it received 241 miles of right-of-way. In return, Litel promised to provide capacity along the system for use by the turnpike for a variety of functions, including warning motorists electronically of changes in weather and road conditions.

Litel also said that SAT was going to act as a later supplier for its electronics and that NEC America was going to provide the initial high-speed electronics. This exemplified the intense competition among the electronics suppliers during this time, with NEC and Fujitsu showing the trump card—workable high-speed electronics—time after time.

AT&T was not the only supplier to bow to the Japanese. The U.S. company Telco Systems was to provide its 560 Mbps electronics to a host of suppliers—including Bell Atlantic and Pacific Telesis. Both ended up using NEC America equipment on routes on which they originally had intended to use Telco Systems electronics. Electra also announced that it was going with Telco equipment, only to have to shift to the Japanese as well when it became apparent the product was not going to be ready in time.

Ericsson also announced that its 565 Mbps electronics was ready, caused LDX to back out of a deal with NEC to buy it, and then lost the contract back to Fujitsu when it could not deliver. Electra also entertained the thought of using Ericsson electronics but could not suffer the delays.

Plessey Stromberg-Carlson announced in autumn 1984 that it received a $2 million order from United Telephone of Florida to provide 565 Mbps electronics, a job that it did complete—although some time later.

It was impossible for carriers not to get caught up in the race to rewire the nation with fiber optics. (See Figure 8.) This was true even of AT&T, whose communications engineers had been accustomed to gradually implementing a new technology without being spurred by the heat and noise of competition.

AT&T in November 1984 announced that it would have a nationwide fiber optic network in place by the end of 1988, including 10,250 miles of optical fiber cable. Specifically, AT&T announced the following fiber optic long-haul routes to be built in the 1985–1988 time frame: Philadelphia to Chicago; Dallas to Houston; Moseley, Virginia, to Atlanta (in effect extending the

Figure 8. Racing to rewire the United States with optical fiber cable was hard work. When large equipment could not be used to lay the cable, crews had to resort to hand trenching. (Photo courtesy US Sprint.)

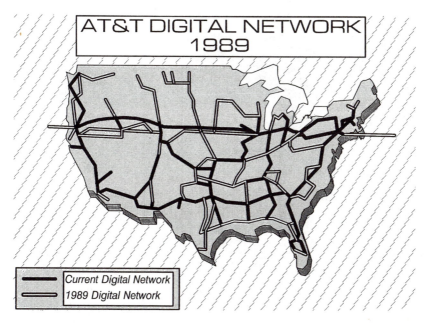

Figure 9. AT&T intends to have nearly 25,000 miles of optical fiber cable in the ground by the end of the decade. This will rapidly accelerate its capabilities to provide digital services. (Map courtesy AT&T.)

Northeast Corridor from Cambridge, Massachusetts, to Atlanta); Anaheim, California, to Tucson, Arizona; Tucson to Dallas; Dallas to Atlanta; St. Louis to Dallas; Chicago to St. Louis; Atlanta to Miami; South Bend, Indiana, to Indianapolis; Nashville, Tennessee, to Birmingham, Alabama; and Nashville to Indianapolis.

However, as the race to be the first to offer coast-to-coast fiber optic connectivity heated up, AT&T accelerated its plans by two years. The 10,250 miles of optical fiber cable were in the ground by the end of 1986. AT&T was promising a massive digital network by the close of the decade. (See Figure 9.) Not surprisingly, MCI was thinking along the same lines, (see Figure 10.)

There was also the realization that not all of the fiber optic networks were going to be built as planned. There would simply be too much capacity. It was estimated, for example, that if all of

Figure 10. Like AT&T, MCI is also working hard to have nationwide digital capabilities. Both carriers are relying heavily on fiber optics in order to accomplish that. (Graphic courtesy MCI).

the fiber optic routes were eventually built, seven billion circuit miles would be created, more than five times the existing number in the United States.

Consequently, consolidation began to occur. The first substantive example of this was a deal between Lightnet and US Telecom. US Telecom had diligently been going it alone in its quest to build a 23,000-mile nationwide fiber optic network. It had now signed up five railroads (the fifth was Chicago Pacific Corporation), giving it access to a total of 15,000 miles of right-of-way.

Lightnet, meanwhile, was aggressively building its own fiber optic network throughout the eastern United States.

The resulting three-tiered memorandum of understanding between the two groups called for United Telecom to buy up fiber optic capacity along Lightnet's routes, in effect becoming Lightnet's first customer. But United Telecom would also be constructing some of the routes and maintaining and operating them. While the construction was generally accomplished as had been mandated, problems eventually arose about who should operate and maintain the network. Suit was brought against Lightnet. The affair was settled amicably out of court in early 1987.

As a further acceptance of and reaction to the situation, Cable & Wireless let it be known that it would not be building a nationwide network—as had been rumored—but had decided to buy and lease fiber pairs from the various carriers that were already building fiber optic networks. The company was therefore scrapping its plans to go into other ventures besides Electra.

Consolidation also had the advantage of providing carriers with capacity to areas beyond the networks they owned. Whatever the rationale, some of the ventures were already finding that they were beginning to enter into the second phase of operations—marketing the networks they were so aggressively building.

Microtel announced that it had already signed up 10 other common carriers by the end of 1984, and Lightnet had signed up the Florida based Americall. By early 1987, Microtel was able to brag that more than 89% of its telephone calls were transmitted digitally and more than 83% used fiber optic facilities.

But the biggest news in consolidation in the fiber optics industry had yet to be revealed. It involved the major regional ventures and was to be known as the National Telecommunications Network (NTN).

On February 4, 1985, NTN formally came into existence. It included four already announced regional partners—Microtel, LDX Net, Litel, and SouthernNet. A fifth partner, Southland Fibernet, had not installed any fiber at the time of the announcement but was going to provide a key fiber optic link between Microtel and LDX Net. The idea, of course, was to join all of the networks together so that more customers could be served.

"The media have raised questions regarding the adequacy of funding, of cable availability, and of unified services for end users," said John Puente, the first NTN chairman. "The NTN pools the resources of five significant fiber optics companies to preempt such concern and provide a unified approach to serving customers.

In establishing the nationwide regional fiber optic network, it was crucial to maintain the integrity of each regional project, explained Puente. The regional networks were moving along ambitiously, spurred by competition, and NTN was going to make sure that was going to continue.

Therein lay the rub, as far as NTN was concerned: how to maintain the growth of each of these bright new stars and form them into a cohesive universe at the same time? It was an interesting challenge—and there were many the new organization had to meet. A second concern, also not so easy, was how to make the venture national in scope (as its name implied) when the new consolidation only went as far west as Texas.

Yet another question had to do with whether individual members should maintain the right to cut outside deals by themselves. LDX Net, for example, was in negotiations with Electra to obtain capacity to western Texas cities. SouthernNet and RCI had also talked about cutting a deal.

At any rate, NTN had reasonable amounts of optical fiber cable already in the ground by the time the project took off and very aggressive plans to continue pursuing installation into 1985 and beyond. (See Figure 11.)

One thorny problem the various carriers had to deal with was how to gain entry into the metropolitan areas they intended to serve, while also maintaining the overall integrity of the fiber optics network they were building. For example, you could have pure fiber optics from Chicago all the way to Manhattan, but if those calls were in turn interconnected on equipment that was 30 years old in New York City, the signal would be diluted.

If you could get the entrenched telephone company to provide fiber optic capacity out to your termination point at a reasonable cost, you were in luck. LDX went this route with Southwestern Bell, allowing it at first to gain entry into major Texas cities. Shortly thereafter, LDX revealed that it was going to pay South-

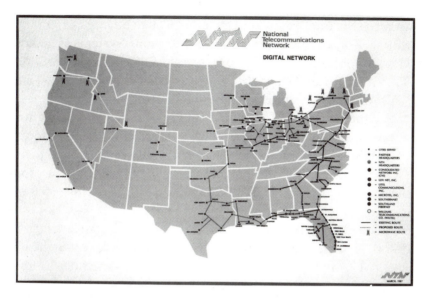

Figure 11. A consortium of seven regional carriers, the National Tele-communications Network had completed 37% of its nationwide fiber optic network by the close of 1985. (Courtesy NTN.)

western Bell to gain access to fiber optic capacity in metropolitan areas in Missouri. The deal provided extensive networking around St. Louis and Kansas City. LDX Net's aggressive con-struction program allowed it to finish its Dallas–Houston link by May, a task completed in less than six months.

Litel, which estimated it would spend $45 million on its fiber optic network in 1985, announced that it had procured full fund-ing for its network, which had been upped to 1600 miles. As part of the effort, Litel received $12.3 million from a consortium of seven French banks. Litel was destined to cut over its first link in April 1985, between Cleveland and Akron, Ohio.

In the early stages, Microtel was making believers of those who wondered if these little ventures could compete effectively against the likes of AT&T. For one thing, Microtel was already building and operating networks. Its Miami–Jacksonville–Tampa link was already cut over.

Early Microtel customers included Teltec Savings Communications Corporation, Allnet Communications Systems Inc., Argo Communications Corporation, Network I, Telesystems, Inc., North American Telephone, Datel Communications, Inc., Transcall America, Execulines, and Microtel itself.

Having written off the failed 432 Mbps electronics it had announced two years earlier, AT&T was now attempting to sell its 417 Mbps electronics, which it said could be upgraded to 1.7 Gbps. Having missed the window so ably covered by the Japanese, AT&T acknowledged its marketing efforts for the electronics were going to be tough.

Like LDX, MCI had also found out about the advantages of doing business with the RHCs. By February 1985, MCI already had at least five deals in the works, some providing fiber-to-fiber interconnect. MCI continued its bullish overall networking plans, aiming to install and operate more than 2500 miles of optical fiber cable by the end of the year.

If Ericsson had problems with its high-speed electronics, it was compensating as an optical fiber cable supplier. It was a major cabler for LDX Net, for example, despite having blown the electronics contract. It also announced a $15 million deal to provide optical fiber cable to GTE Sprint.

It was no wonder AT&T was finding rough sledding trying to market its 417 Mbps electronics. Many suppliers were already aiming at the next higher rung on the digital ladder (560 to 565 Mbps). Fujitsu had even begun talking about supplying an 810 Mbps system—but that wouldn't happen until early 1986.

In the meantime, a host of suppliers were caught in bloody battle trying to get their 565 Mbps electronics out first. To further fuel the fire, front-runners Telco Systems and Ericsson had squandered the leads they said they had, leaving the competition wide open.

There were already 14 would-be high-speed electronics suppliers by June 1985 who were aiming to break through the 405 Mbps level and eventually move on to higher rates. They included: Fujitsu, NEC, Telco Systems, Rockwell-Collins, Northern Telecom, AT&T, N. V. Philips, SAT, CIT-Alcatel, Plessey, Siemens, NKT of Copenhagen, Ericsson, and Standard Telephones and Cables.[4]

The safest bet for using fiber optic electronics in the minds of many carriers however, continued to be the Japanese. RCI, for example, said it would be using Fujitsu as its supplier. AT&T received the contract to provide the cable. Fujitsu also announced that it was providing the 400 Mbps electronics for NTN member Southland Fibernet.

Surprisingly, it was AT&T who was using the first 565 Mbps electronics, and it was being manufactured by the Dutch based N. V. Philips. The first link in the United States to use the electronics was AT&T's route between Poughkeepsie and White Plains, New York[5] Other early AT&T routes included Providence–Framingham, Massachusetts (43 miles), Manchester, New Hampshire–Littleton, Massachusetts (53 miles), Detroit–Toledo (60 miles), and Detroit–Plymouth, Michigan (22 miles).

Despite its problems, Telco Systems was to find an early happy customer in Teleport Communications.

Rockwell-Collins also claimed early success in providing its 560 Mbps electronics. Rockwell-Collins was destined to become the first non-Japanese supplier of high-speed electronics to MCI. Rockwell International of Canada later was to provide fiber optic electronics to Telecom Canada's extensive fiber optic system.

Perhaps using the edge provided by its deal with subsidiary United Telephone of Florida, Plessey Stromberg-Carlson scored a three-year contract to provide its 565 Mbps electronics to US Telecom, a volume purchase agreement estimated to be worth $35 to $45 million.

While the Plessey effort incorporated basic technology developed by the British company, it was a team of Americans headed

---

[4]"Coming to Market With the 565s and Beyond," *Fiber Optics News,* June 24, 1985.

[5]"Philips Cuts Over 565 Mbps Link for AT&T Communications," *Fiber Optics News,* July 22, 1985.

by engineering director Whit Cotten who developed the product for the U.S. marketplace. This was accomplished by developing the M34E muldem that was able to combine three DS3 digital signals into a 140 Mbps stream, the European standard, and vice versa.

US Telecom also hung tough with Ericsson electronics, even through the period that Ericsson was having troubles, and Ericsson's high-speed electronics eventually made it into the US Telecom network.

Despite the fact that it had only been announced months before, NTN soon added two additional members. They included Consolidated Network, Inc., and Williams Pipeline subsidiary Williams Telecommunications (WilTel). Consolidated Net was filling a key gap between St. Louis and Indianapolis, and initially planned to build a 240-mile network.

It was WilTel, however, that promised to give NTN a dimension it had previously lacked. WilTel announced that it would build a fiber optic network between Omaha, Nebraska, and the West Coast, thereby giving added credence to NTN's advertisement of a national network. This was later modified so that the route connected Kansas City with Los Angeles. (See Table 1)

WilTel also announced it intended to build a digital microwave system from Salt Lake City to the Pacific Northwest, interconnecting Portland, Oregon, and Seattle, Washington. All this was in addition to a 1200-mile network it intended to build between Kansas City and Des Moines, Iowa. Much of WilTel's installation was to occur in oil and gas pipelines owned by parent Williams Pipeline Inc.

Meanwhile, consolidation among the carriers continued to increase. Lightnet scored impressive gains, signing Cable & Wireless and RCI as customers. Cable & Wireless was going to use capacity along Lightnet's Washington–Chicago route, while RCI intended to use Lightnet capacity along its New York City–Washington, D.C., corridor.

As a further example of consolidating the telecommunications networks, MCI announced that it was purchasing Satellite Busi-

Table 1. By the close of 1985, the National Telecommunication Network had already made significant progress. The largest individual partner and the last to join NTN was Williams Telecommunications. While Wiltel had just begun efforts when NTN issued this status report, Wiltel was to eventually assure the nationwide capacity NTN's name promised.

Company	Route Miles		Miles Operational	Planned Cities	Cities Operational
	Planned	Completed			
Consolidated Net	731	285	0	13	0
LDX Net	2,100	1,240	1,012	20	18
Litel	1,561	905	675	27	9
Microtel	1,294	663	556	33	15
Southernnet	1,250	376	105	24	2
Southland Fibernet	332	236	69	6	2
Williams Telecom	3,571	304	0	18	0
Total	10,839	4,009	2,417	141	46
Percentage		37%	22%		33%

* Source: National Telecommunications Network

ness Systems from IBM in a financing arrangement that left IBM with an 18% chunk of MCI stock. The two networks were to be consolidated, including fiber optic capacity that SBS had already purchased. MCI was also to later purchase General Electric's RCA Global Communications, Inc.

The first casualty among the announced fiber optic carriers was Fibertrak. The venture had yet to install an inch of cable by July 1985 and announced that all of its construction plans were being delayed indefinitely. Fibertrak never did get going.

GTE Sprint and US Telecom announced that they were going to be swapping fiber optic capacity. This was to be a harbinger of the eventual merging of US Telecom and Sprint into US Sprint.

One new venture, Indiana Switch, was to be 50% owned by US Switch and 50% by 27 independent telephone companies in

Indiana. Indiana Switch planned to build a 600-mile fiber optic network in Indiana at the preliminary cost of $19 million. Similar ventures were also being considered in other states, including Illinois, Michigan, and Wisconsin.

Attracted by the boom the rewiring of the American countryside was bringing, additional suppliers tried to gain entry. Two British based cablers, BICC and STC, scored orders in the United States. Both were to supply US Telecom.

This massive undertaking—rewiring America's countryside with fiber optics—did not go unnoticed by major communications players. Speaking before a Senate panel chaired by former Senator Barry Goldwater (Republican-Arizona), then FCC Chairman Mark Fowler cited the fiber optics activity as evidence of a free market system at work. Then AT&T Chairman Charles Brown characterized the phenomenon as evidence of long-distance marketplace competition and calculated that roughly $3.6 billion would be spent on the rewiring by the end of the 1980s, resulting in 3.1 billion circuit miles of new capacity.

Meanwhile, several of the carriers announced that they were fully financed and were going to complete what they had said they were going to. LDX Net claimed that it did not require additional financing to complete its network, and US Telecom said that it was going to make it, even though earlier it had looked as though extra financing was going to be required. US Telecom said that a switching deal it negotiated with Northern Telecom was its final step toward network completion.

On and on the announcements came, almost in an endless progression. As yet another example of interest, the utilities were beginning to take seriously the promise of fiber optics.

In October 1985, five midwestern utilities announced they were banding together to create five subsidiaries that, in turn, were going to build a 650-mile fiber optic network to be known as Norlight. The utilities included Wisconsin Public Service Corporation, Dairyland Power, Wisconsin Power & Light, Madison Gas and Electric Company, and Minnesota Power.

To accommodate the rising new utility market, Ericsson joined forces with Reynolds Metals to provide optical power grounding wire (OPGW) for utilities. Sumitomo later joined with Alcan Aluminum Company to manufacture OPGW. Kaiser Aluminum & Chemical Corporation joined together with Siecor. Advantages of OPGW were that it carried communications along aerial lines and could also ground out power wires.

The level of intensity was reaching the breaking point as 1985 came to a close. Thousands of miles of optical fiber cable were being installed and even more were planned as the carriers roared into 1986.

Nineteen eighty-six was to be the year that an incredible amount of optical fiber cable was to go in the ground. It will be remembered as the leading year optical fiber cable went into the nation's countryside. It will also be remembered as the first year a coast-to-coast fiber optic call was made and intense competition revolved around the right to claim that honor.

First of all, there was US Telecom, which was now in the process of joining GTE Sprint to become US Sprint. This ambitious carrier was aiming to install 10,000 miles of optical fiber cable in one year, a herculean effort if it could be done. (See Figure 12.)

NTN was also beginning to interconnect its partners, an essential step in forming the national loop it hoped to create. The first interconnections involved Microtel and Southland Fiber Net in Tallahassee, Florida, and LDX Net and CNI in St. Louis, Missouri. More interconnections followed.

Having scrapped their more conservative plans to increase installation incrementally, AT&T and MCI also joined the contest to be the first to offer coast-to-coast fiber, a competition that was also to be promoted heavily in the media with sometimes misleading advertising.

This wholesale construction was having an impact in real dollars. The Commerce Department predicted that 1986 would be the year the fiber optics industry topped the $1 billion mark in the United States, constituting a $1.1 billion market. KMI esti-

Figure 12. Workmen bury optical fiber cable in a just completed ditch. US Sprint intends to have 23,000 miles of cable installed by the time its network is completed. (Photo courtesy US Sprint.)

mated that a total of 638,000 fiber km was installed in the long-haul network in 1986, representing 56.06% of the total market. (See Figure 13.)

While the industry continued to grow, it also continued to consolidate. Once thought to be competitors, Litel and RCI agreed to become customers of each other. Litel, through RCI, now had access between Cleveland and Buffalo, New York, between Buffalo and New York City via microwave, and between New York City and Washington, D.C.

The earlier negotiations between SouthernNet and RCI paved the way for a joint agreement between that NTN member and RCI. SouthernNet made it clear, however, that it was acting in its own interests and not those of NTN. RCI had earned much respect in the fiber optics and telecommunications industry, having completed its Buffalo–Chicago build in less than one year.

SouthernNet was to cut over its Richmond–Charlottesville route in spring 1986.

The inevitable question arose as to when an NTN member should act in its own behalf, as opposed to when it should act as an NTN member. NTN President Lawrence McLernon, who had been elected to lead the venture in 1986, set policy that NTN members were free to interconnect with any company, although McLernon requested that they use other NTN members where possible.

Perhaps the most extreme example of this was when NTN member LDX Net announced that it was going to provide its own nationwide capacity in addition to that offered over NTN. While relying mainly on NTN, LDX Net sought to forge its own

Figure 13. The year 1986 will be remembered as the year that more optical fiber was used in the long-distance market than at any other time. The optical fiber is marked and stored at AT&T's Atlanta Works facility. (Photo courtesy AT&T.)

alliances with other fiber optic carriers, such as Lightnet. As rationale for the decision, the carrier claimed it was much easier to sell the networking when it said it would be responsible for maintaining service, rather than saying it would have to coordinate with up to six other entities in case a problem arose, according to Don Hutchins, who was promoted to president of LDX Net in December 1986.

NTN was still grappling with the question of how strong its central organization should be. Started under a "strong partner/weak central union" formula analogous to America under the Articles of Confederation, NTN was having trouble growing into a strong union. This was typified by the fact that its central staff remained small, while the joint revenues of the various partners exceeded $1 billion. NTN did start up a nationwide marketing effort as 1987 began, but LDX had already begun an aggressive effort.

It was becoming clear that long-distance fiber optic carriers were entering a new phase. With the realization that routes were being built as planned also came the understanding that the capacity along them was going to have to be filled.

NTN and AT&T began offering 45 Mbps fiber optic service to potential customers, through which the customer could purchase that amount of capacity. AT&T was offering the Vivid System, a 45 Mbps full-color, full-motion switched video system over fiber. AT&T was also to offer 1.5 Mbps fiber-only service. Microtel brought on a new president, Dale Gregory, who had marketing savvy. New Lightnet President Arthur Parsons also stressed the importance of selling the company's capacity. Lightnet was offering DS-2 service by the summer of 1986.

Companies who used one or more of these carriers for capacity and then helped the customer interconnect with the network were also making inroads. These so-called resellers often provided service in smaller chunks than the carriers could, attracting a larger base of customers, and claimed that reduced overhead kept their prices below those of the carriers.

These marketing efforts often meant stepped up discussions with large corporations, who were increasingly trying to design their own networks. Electronic Data Systems was in the midst of designing a nationwide network for General Motors that was to incorporate significant amounts of fiber optics, with EDS using Sprint to provide large amounts of its capacity. Even U.S. satellite carrier Comsat was looking for terrestrial capacity that would be largely fiber based.

The need on the part of the major carriers to provide higher data rates persisted. MCI announced at its annual meeting in May 1986 that it had cut over Fujitsu electronics operating at 810 Mbps between Manhattan and White Plains, New York, the first commercial use of electronics operating at that speed. NEC said it had reached the 1.1 Gbps level in experimental tests, and Pacific Bell was testing the electronics in the field by the end of 1986. NTT was experimenting with electronics operating at 1.6 Gbps, and AT&T intended to introduce 1.7 Gbps into commercial use sometime in 1987.

Pacific Bell introduced a commercial 1.12 Gbps system at the beginning of 1987 that used NEC electronics over a 2.5-mile stretch between two of its central switching offices in San Diego.

As rewiring the countryside proceeded, innovative ways of gaining access to right-of-way also continued. MCI struck a deal with the California Department of Water Resources to install fiber optics along the state's aqueduct. The subject of opening up federal highways to the installation of fiber optics became the topic of a Congressional hearing, and the Department of Transportation agreed to study regulations that prohibited its use.

Other ventures also recorded gains. Indiana Switch received approval from the FCC to launch into action, despite strong protests from AT&T and MCI.

But the real story continued to be the race to see who would be the first to rewire America from coast to coast with optical fiber cable. The four carriers involved were finding that fiber optics as a technology had sales potential, and being the first to institute a nationwide fiber optic network would be a feat long remembered. (See Figure 14.)

To show its commitment to fiber optics, US Sprint launched a $70 million advertising campaign whose theme was that "every fiber optic call sounds like you're right next door." Sprint claimed that independent tests had proved that "three out of five business people preferred the clarity of US Sprint over the typical noise of AT&T." (This was later to be upgraded to 7 out of 10, with 90% of those saying they would switch from AT&T because of it.) Sprint ignored mention of the other two contenders, MCI and NTN, in its ads.

Figure 14. Tens of thousands of miles of fiber cable were laid in 1986 as the race to provide the first coast-to-coast call was on. (Photo courtesy US Sprint.)

Sprint clearly was "leading the charge of the light brigade," *The Wall Street Journal* reported in September 1986.[6] Sprint President Charles Skibo maintained the carrier was establishing a new "platinum standard," replacing AT&T's tarnished "gold standard." (Skibo has since left the company.)

The Sprint campaign was wildly successful, Sprint reported, with 250,000 new business and residential customers coming on board the first month (July 1986) US Sprint was in operation. That number was to soar to two million by the end of 1986. "Our success is attributed to the public's response to our modern fiber optic network," said John Birk, president of Sprint's Northeast Division.

Sprint aimed to have two West Coast builds completed by the end of the year, one joining Chicago with Cheyenne, Wyoming, and San Francisco, and a second joining Fort Worth, Texas, with Phoenix and San Francisco. (See Figure 15.) More than 300 work crews were in the field daily installing fiber as Sprint raced to nationwide connectivity.

Sprint's competitors were not exactly rolling over and playing dead, however. In autumn 1986, MCI announced that it would soon be cutting over coast-to-coast fiber and launched a belated advertising campaign of its own, in which it announced that it had "pioneered single-mode fiber optic technology." In December, MCI announced that it was only 30 miles away from nationwide fiber optic connectivity. But MCI acknowledged it was that close only because NTN member WilTel was that near to completing its network. MCI had cut a deal with WilTel to lease capacity from it. (See Figure 16.)

MCI also announced that it would be building its own fiber optic network going from Chicago to Dallas to Phoenix to Los Angeles using right-of-way provided by Burlington Northern, but that route would not be completed until later in 1987. MCI

[6]"Fiber Optics Promises High-Tech Revolution," *The Wall Street Journal*, September 9, 1986.

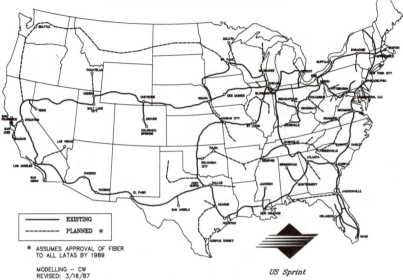

Figure 15. Sprint's plan was eventually to have three routes to the West Coast. (Map courtesy US Sprint.)

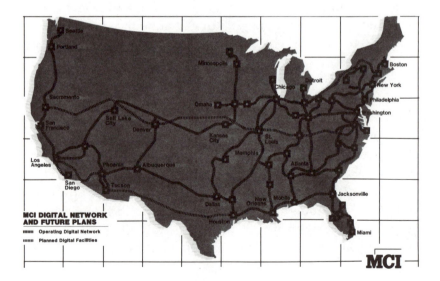

Figure 16. MCI's operating and planned nationwide fiber optic and microwave route. (Map courtesy MCI.)

sources claimed a 15% staff cutback would not impact the fiber optic effort.

As of November 30, 1986, NTN claimed that it had completed 8652 miles of its planned 11,951 mile network. But it became clear that NTN was stuck in the same boat as MCI—waiting for the WilTel build to be completed.

Several carriers advertised that they would be providing the first coast-to-coast fiber capability. Two included US Sprint (See Figure 17.) and MCI. (See Figure 18.) Both claims turned out to be premature.

AT&T's accelerated schedule left the carrier claiming that it would have coast-to-coast capabilities by the end of 1986. Like MCI, AT&T suffered a reduction in staff in 1986; also like MCI, AT&T claimed that its fiber optic activities would in no way be diminished.

Without fanfare, AT&T cut over the first commercial fiber optic coast-to-coast call on December 5, 1986. The call was placed between Boston and Los Angeles, and AT&T continued transmitting coast-to-coast optically from that day forward, according to spokesperson Rick Brayall.

Sprint announced its first transcontinental fiber optic transmission on December 31, 1986. Skibo placed the call from New York City to Los Angeles. (For its first transcontinental video-conference, also on New Year's Eve. Sprint reunited a U.S. Marines recruiter in New Jersey who could not get home for the holidays with his family in California.)

Of Sprint's planned 23,000 mile network, nearly 15,000 miles were completed by the end of 1986, with 8,480 miles of that operational. "We will have our fiber network virtually completed by the end of 1987, and at that time 97% of our traffic will be wholly fiber," Skibo said. "Just as important to US Sprint customers, 97% of the telephones in the United States will be accessible to our fiber optic network." US Sprint's commitment to fiber optics was so strong that Sprint was later to place its entire old telephone system on the auction block. The move exemplifies Sprints' commitment to fiber optics.

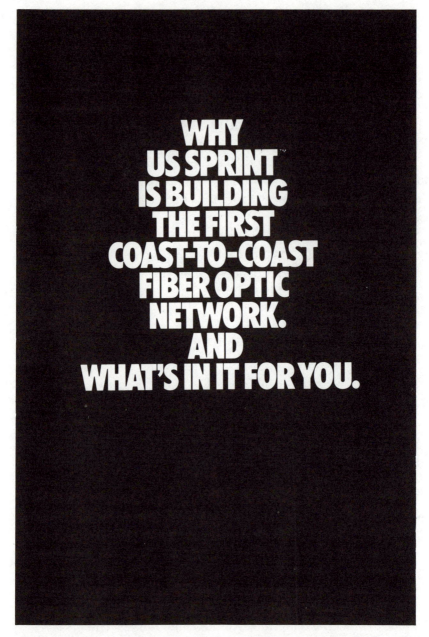

Figure 17. MCI and US Sprint both launched media campaigns to tell the world and their customers that they were coming out first in the struggle to rewire America. Sprint took out a number of full page ads; this one appeared in *The Wall Street Journal*.

Starting in January, The Information Age comes of age.

MCI® ANNOUNCES THE FIRST COAST-TO-COAST FIBER OPTIC NETWORK.

Figure 18. This MCI ad, appearing in the December 30, 1986 *Washington Post*, came one day before Sprint made its first coast-to-coast fiber optic call, which was weeks prior to MCI's announcement of its first call. Both were bested by AT&T, however.

WilTel did complete its build to the West Coast by mid-January, in effect providing coast-to-coast capability to three carriers, including NTN, MCI, and LDX.

American businesses and customers were finding out about the promise of fiber optics, as can be demonstrated by the US Sprint story. Selling fiber optics technology, Sprint added two million new customers in six months, serving 475 of the Fortune

500 companies by the end of 1986. Other carriers would continue to brag about how much fiber optics capacity they had. A laboratory curiosity two decades earlier, fiber optics had made its impact throughout America and was to continue to grow in importance.

There was poetic justice in the fact that AT&T was to reach the milestone first. From the beginning, the company had spent enormous amounts of time and money in creating fiber optics and watching it grow. Certainly the carrier had suffered bumps and bruises along the way, but it had never wavered in its commitment to the technology.

AT&T was also exhibiting strong staying power, announcing in early 1987 that it intended to install 13,000 new domestic optical fiber cable miles between 1987 and 1989, bringing its total count to almost 25,000 miles by the end of the decade. This second-tiered strategy was geared to connecting large numbers of additional population centers into the company's nationwide fiber optic trunk. Some of the cities to be included were Milwaukee, San Diego, Little Rock, New Orleans, Seattle, Orlando, Denver, Minneapolis, and Buffalo.This "domestic" system includes an underwater fiber network between the U.S. mainland and Puerto Rico.

AT&T's attitude was in marked contrast to other long-distance carriers, who were winding down significantly in 1987.

The rewiring of America's countryside was becoming a reality. An agreement between Sprint and Telecom Canada to build a 324-mile fiber optic network between Springfield, Massachusetts, and Montreal, Canada, demonstrated that the phenomenon was spilling over international boundaries as well.

But fiber optic suppliers were also eyeing another new market, something even closer to home—the cities of America.

Chapter 8

Rewiring the Cities

I F IT SEEMED as though the long-distance companies were putting tremendous amounts of optical fiber cable in the ground, the Regional Holding Companies carved out as part of divestiture were doing even more. Some of the credit for that goes to AT&T, which had acquainted them with fiber optics when they were all one big happy family.

By the time divestiture came into effect, however, the Regional Holding Companies saw fiber optics not as some technological marvel that was nice to have, but as an economic necessity to help battle the threat of other carriers coming into their precious local exchanges and claiming some of the business for their own. Bypass carriers have found out about fiber optics, and it is expected to be the dominant transmission medium for them.

The Regional Holding Companies (RHCs) were free to use fiber optics in their service areas (LATAs-local access and transport areas). While those areas encompassed the major metropolitan areas, some also spread farther out into the country. (One LATA, for example, encompasses the entire state of Utah.) Consequently, the RHCs are installing fiber in the cities, but their job is also to install it in metropolitan areas and sometimes in the surrounding communities, as well. (See Figure 1.)

Telephone Outside Plant Network

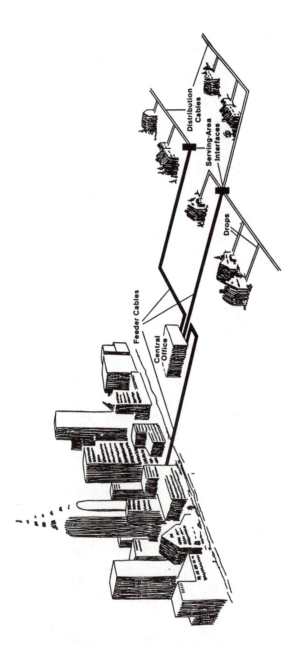

Figure 1. The job of the telephone companies and local exchange carriers is to bring telecommunications services to homes in and near metropolitan areas. This represented a second frontier for the development of fiber optics, following the rewiring of the countryside. (Diagram courtesy Corning Glass Works.)

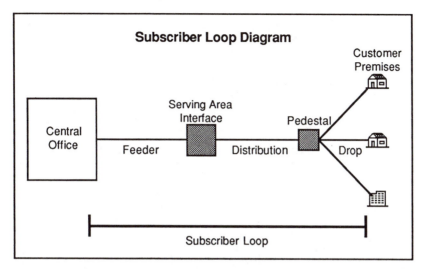

Figure 2. The future large-scale development of fiber optics is expected to be in the subscriber loop, which consists of the feeder, distribution, and drop segments. (Graphic courtesy AM Consulting Inc.)

The non–long-haul market consists of intra-LATA applications by the Bell Operating Companies and the independent telephone companies. That includes the interexchange market and the local exchange. The interexchange market consists of a telephone company connecting its telephone exchanges together. The subscriber loop is the local traffic carried within each of those exchanges. It includes feeder, distribution, and drop segments. The future large-scale growth of fiber optics is expected to be in the subscriber loop. (See Figure 2.)

Robert Bowman, who left Telco Systems to head Artel Communications, characterized the local exchange has as "the next frontier."

The same economics driving the long-distance companies also drove the RHCs at the beginning. Fiber costed in for the RHCs for their longer distance applications such as interexchange applications. But the telephone companies have also brought fiber optics into the greater metropolitan exchange and into the downtown areas, albeit at a slower rate.

Table 1. The use of fiber by the Regional Holding Companies increased significantly from 1984 to 1985. This included the amount of fiber that went into the loop. (Source: Siecor Corp. *)

Telephone Company Fiber Applications				
(Figures in thousands of fiber km)				
	1984		1985	
Regional Holding Company	Total	Loop	Total	Loop
Nynex	40	15	75	25
Bell Atlantic	35	15	55	25
BellSouth	55	35	90	30
Ameritech	40	15	75	35
SW Bell	15	5	65	15
U S West	20	5	40	10
Pacific Tel.	35	15	65	25
Independents	10	—	15	5
	250	105	480	170

*Information in the enclosed chart are estimates only. Siecor makes no warranty as to their accuracy. Siecor further makes no representation as to whether consent of the listed companies may also be required.

In 1984, the first year following divestiture, the RHCs and independent telephone companies installed 250,000 fiber kilometers, according to Siecor Corporation.[1] The following year that number almost doubled, to 480,000 fiber km. (See Table 1.)

Table 1 reveals the bullish attitude that BellSouth has taken toward fiber optics. Contributing factors to BellSouth's heightened awareness about fiber optics may have spawned from the fact that AT&T's first commercial prototype experiment was in Atlanta. It is also interesting to note that the largest two cablers—AT&T and Siecor—are both located in BellSouth's service area, although Siecor's Hickory plant is served by Centel.

All of the Bell Operating Companies will use fiber as an integral part of their growth and ISDN implementation strategies, the Siecor study said. Motivating factors include: the critical

---

[1]"Siecor Recommends Fiber-Based Universal Network," *Fiber Optics News,* December 23, 1985.

need for new facilities to accommodate growth; expansion of revenues through new services; the need to respond to the threat of bypass; and ongoing efforts to improve operating efficiencies by lowering maintenance and administration costs.

One of BellSouth's two operating companies, Southern Bell (the other is South Central Bell), set the OFC '85 seminar abuzz with the news that it hoped to soon have an integrated all-fiber digital switched video loop for a residential community, bringing fiber to the home for the first time.

Southern Bell's Richard Snelling explained that a large research effort is required that would include manufacturing and materials development for a cost-effective integrated loop fiber facility. Techniques for volume manufacturing of optical components and their integration with electrical components still needed to be devised, however, he said.

After one false start, the so-called Arvida project, BellSouth found a project in Florida that it is wiring all the way to the house. It is the Hunter's Creek project. Hunter's Creek had 30 families wired with optical fiber by the end of 1986, with homes continuing to be fibered as they are built. Genstar Development Inc. is the cable company operating the system.

But RHCs besides BellSouth have been going fiber and have not been afraid to tell the world about it. Bell Atlantic Marketing Vice President William Newport told analysts in 1984 that the carrier was ringing Pittsburgh, Philadelphia, cities in Northern New Jersey, Washington, D.C., and Wilmington, Delaware, with fiber.

"It's clear that optical fiber is going to be very big in our future," Newport said. "We know that customers want the flexibility it delivers, so we're investing heavily in that technology to provide competitively priced services that enhance the value of our local network."

Ameritech spent approximately twice as much in 1984 on fiber optic installations as it did the previous year, according to Ameritech Development Corporation President Jim Bauer. Expenditures increased from $23.2 million in 1983 to $45.6 million in

1984. Fiber optics "is Ameritech's medium of choice," said Bauer.

Bauer cited the fact that subsidiary Illinois Bell was part of the first commercial AT&T installation of fiber optics—the Chicago project.

While accounting for only a small share of the market, the independent telephone companies were also being drawn to fiber optics. Small independents such as Lincoln Telephone Company, Lincoln, Montana, Cumby Telephone Cooperative in Texas, and Nebraska Central Telephone were making the decision to install optical fiber cable, even though none of these had what could be considered a large capital expenditure budget.

One basic task the RHCs had after divestiture was to entice the long-distance carriers into interconnecting with them and not with a bypassing competitor. In order to do that, an RHC sometimes had to build a fiber optic link to the edge of its prescribed service area so the long-distance carrier could interconnect to it fiber to fiber.

AT&T, of course, had been providing its long-distance service to the RHCs when they were one organization and continued to do so. Other carriers such as MCI, however, were also working things that way when necessary. By early 1986, MCI had consummated deals for fiber-to-fiber interconnect with Pacific Bell, Mountain Bell, Bell of Pennsylvania, Southwestern Bell, New Jersey Bell, and Ohio Bell.

The National Telecommunications Network was also using the RHCs for a good portion of its entry into metropolitan areas and for other specialized work. LDX early on announced that Southwestern Bell was providing it access to various cities via fiber. Mountain Bell was paid $5.26 million to build a 121.5-mile network in the Rocky Mountains for Williams Telecommunications as Wiltel headed to the West Coast.

Sometimes the RHCs had no choice but to build fiber, due to deficiencies other transmission mediums manifested. When a 40-story building was constructed in Kansas City, in effect knocking out its microwave path, Southwestern Bell installed a 41-mile fiber optic system between its offices in Kansas City and Lansing, Kansas.

In addition to its "Ring around Manhattan," NYNEX was also doing other ambitious fiber jobs. NYNEX received a reported $4.6 million to build a private fiber optic network for Digital Equipment Corporation. It also planned the Interborough Optical Network to connect all the boroughs in New York, as well as adjacent areas. Also included was a "Ring around Westchester" and "Under the Bridge," a plan to connect Staten Island with Brooklyn via fiber under the Verrazano Bridge.

NYNEX also intended to build "Video around Manhattan", providing major Manhattan customers with video services. NYNEX also expanded the use of fiber optics on Long Island and was slated to add 2000 fiber miles to existing trunk. NYNEX also was instrumental in stringing fiber from Manhattan to Liberty and Ellis islands for coverage of the Statue of Liberty centennial celebration.

New England Telephone in 1986 began pumping $1 million into expanding the downtown Boston fiber optic network. The telco was hoping to add 200 fiber miles in an admittedly cong-ested area.

Of the components of the subscriber loop detailed in the sub-scriber loop diagram, the most fiber activity was concentrated in the feeder portion, according to Barry Mullinix, who has since left Siecor Corp. to form his own consulting company, AM Con-sulting, Inc.

The feeder loop is well on its way to being fibered, Mullinix says. The economics as of the mid-1980s were right for installing optical fiber there. Of the digital carrier installations that require new cable construction, nearly 100% are going in over fiber. "If I am going to bury new cable, its going to be fiber," Mullinix says.

But what Snelling was referring to in the OFC '85 paper, and what gets marketers such as Mullinix very excited, is the poten-tial use of fiber optics in the distribution and drop sections of the loop. In other words, running optical fiber directly to the custom-tomer's home. The size of that market will be enormous, pre-dicted to be 10 times the size of what it is for the feeder loop.

Mullinix pegs the time when that will happen as the late 1980s or early 1990s. Before that happens, technological refine-ments will have to occur. As Mullinix points out, fiber optic sys-tems initially were designed for long-haul applications, with a basic goal of sending signals long repeaterless distances. The goals in designing fiber optics for the local exchange are some-what different. While Mullinix believes that the basic technology has been developed, fiber optics needs to be tailored for this exciting new marketplace. There has also been increasing pres-sure to bring fiber into the local exchange to keep the industry healthy.

The RHCs continued to exploit aggressively the advantages of fiber in the mid-1980s where they could, using more than 500,000 fiber km in 1985 and more than 400,000 fiber km in 1986. But the decline experienced from 1985 to 1986 continued into 1987, when KMI initially predicted the RHCs would install less than 400,000 fiber km, even though installations from inde-pendent telephone companies continued to grow. Two factors were blamed for the diminishing demand. For one thing, fiber was not costing in rapidly enough in the distribution portion of the loop to pick up the longer distance applications, which were beginning to tail off. Another problem, ironically, was that better electronics were reducing the amount of fiber necessary to do the job. The result is that so-called last mile applications have to start costing in for the industry to fire back up to its previous high.

This trend has begun, as Kessler found that 22% more fiber went into the subscriber loop in 1986 (245,000 fiber km) than in 1985 (202,000 fiber km). Kessler also noted, however, that as fiber gets closer to the home and office, the amount of fiber will diminish compared to long-haul applications but the required electronics will increase.

Fiber optic feeder links that already have begun proving in economically include telephone operating companies serving growing communities, densely populated areas, and cities with high concentrations of large business customers, according to Corning.

"For us, use of fiber in the feeder is strictly an economic issue," explained Win Beller, staff manager of technology and planning for Pacific Bell. "Sophisticated computer analysis takes into account existing facilities, available duct space, cost of money, and many other factors," he said. "To date, our fiber installations have been for relief of existing facilities, when cost analyses determined that fiber would be the most economical choice. Once they're installed, they work great."[2]

Not surprisingly, BellSouth takes a more bullish attitude, with BellSouth operations manager Harry Taylor explaining that such installations are a matter of policy. "Our Sunbelt location gives us an opportunity to spend more capital on growth than some other companies," said Taylor. "We're going to build the network of the future—a fiber optic digital connected network."

"Bell of Pennsylvania's future plans call for aggressive penetration of fiber into the feeder plant," according to Leonard Sather, manager of outside facilities engineering. "We'll use single-mode exclusively. By the beginning of 1987, these fiber installations will be routine."

AT&T has recognized the potential market for and the need to get fiber optics into the local exchange for some time. In the early 1980s, it introduced its Fiber SLC carrier systems, which (like its other SLC 96 services) provided customers with up to 96 voice channels over T1 lines. What AT&T was selling in the early days was potential—the ability to position the customer for dynamic growth.

That dynamic growth was beginning to be realized several years later as AT&T introduced the DDM-1000 Loop Multiplexer. The multiplexer could effectively upgrade the SLC carrier

[2]"Optical Fiber 'Proves In' In The Feeder," *Coming Guidelines,* Vol. 2, No. 3.

system employed from the DS-2 level to the DS-3 level, in effect increasing service capabilities sevenfold while using the same optical fibers.

Corning is also gearing up for the day of "fiber to the home" by trying to develop optical fiber cable that will be easier to handle and work with. "It's so cumbersome to work with today that it is difficult to see how it could be deployed en masse," Mullinix observes. "A major effort is going to be to reduce manufacturing and installation cost. Now anywhere from one-third to one-half of installation costs are with labor."

Taken from that perspective, one cannot help but recall the early decision to use multi-mode fiber for the first commercial applications, with one reason being that it was easier to handle. Mullinix strongly recommends the implementation of single-mode fiber for these installations, but notes that some companies have already begun installing multi-mode fiber for feeder applications to keep costs down.

Getting the economics to conform will be tricky business as the Hi-Ovis, Biarritz, and Elie-Manitoba experiments demonstrated. The Times Fiber lesson is also living proof that disaster awaits those who attempt to tackle this market without the economics on their side to do it.

Snelling characterized the main problem with getting there as hardware. He predicted that cable television companies and telephone companies will merge by 1990. Southern Bell may sell its information centers in the home with a mortgage, a new idea for the franchising process. An average cost could be $1500 per home, with about 69% going for terminal costs.

At OFC '86, Bell Laboratories' Thomas Wood described a system designed by Bell Labs that Wood says has the potential for being low cost. In the setup described, one laser is used to send a signal over a bidirectional fiber to a detector, which also acts as a modulator (known as a multiple quantum well modulator). The fact that a laser is used away from the house obviates any potential safety threat. The Bell Labs system transmitted to distances of 3.3 km.

Plessey also demonstrated a bidirectional link geared for use to the home. The Plessey system also used a single source that could send signals to 2 km.

A switched optical communication system within the subscriber loop could have an initial transmission capacity of 560 Mbps, according to the Siecor study, which was written by Robert Strock, who has since left Siecor. A single fiber could accommodate simultaneous delivery of the following services: 4 channels with the capability of delivering high definition television (100 Mbps per channel); 30 high fidelity audio channels (1.5 Mbps per channel); four narrowband ISDN channels (144 Kbps per channel); and 12 telemetry channels (low bit rate channels).

But the major constraint is still cost. Laser transmitters, for example, would have to come down from $3000 to $200, according to Donald Fye, of GTE Laboratories. Splitters would have to come down to $20 from $70, and connectors would have to drop from $150 to $20.

A laser packaging design that could help to speed the advent of fiber into the local exchange by decreasing the cost of a laser from the thousand dollar range to a fraction of that was announced by BellCore researchers Paul Shumate and Leslie Reith in autumn 1986.

The technique uses technology borrowed from the Japanese relating to compact disk packages, which use lasers very inexpensively. The package also includes a telecommunications laser and specially designed lenses about the size of pencil leads.

While noting that questions still remain, "the cost factor now looks very manageable," says Reith. "We're hopeful of seeing lasers in the local loop a lot sooner than previously expected."

The thought of a massive new marketplace opening up to peoples' homes and offices has led to corporate positioning and hundreds of millions of dollars being spent in research and development to be waiting with the right products.

A number of U.S. companies chipped in to launch an R&D program being run under the auspices of Battelle Laboratories to pool resources to develop devices that could be placed into the local exchange. A major reason for this was to try to counteract the anticipated Japanese penetration.

Corning continued to automate its facilities in Wilmington, North Carolina, so by the end of 1986 it was capable of manufacturing from 2 to 2.3 million fiber kilometers annually. AT&T is rumored to have a similar capacity. The object is the local exchange, since long-haul market installation peaked with the tremendous drive in 1986.

By then, indications of the massive amounts of fiber to be used in the local exchange were already starting to present themselves. Bell Atlantic was using 156 fibers per cable provided by Siecor to install in Washington, D.C. Southwestern Bell installed 144-fiber cable in Kansas City. Fiber pioneer AT&T has unveiled a cable housing 408 optical fibers. (See Figure 3.)

Knowledgeable officials such as AT&T's John Moran were urging companies to wire new buildings with fiber so those structures would have the potential of becoming "smart buildings," buildings that can provide a variety of communications needs to their tenants. That was the case even where companies had to leave the fiber "dark" or unused until the electronics was able to cost in.

Installing optical fiber cable in buildings made sense, considering the large amounts of fiber anticipated to come into the loop that can interconnect with them, according to Moran. An example of tailoring product for these applications is AT&T's optical fiber riser cable, which conforms with the 1987 National Electrical Code.

Fujitsu America announced its first fiber optic products geared for the local exchange in spring 1986. Included was a 135 Mbps optical fiber transmission system, a digital multiplexer-demultiplexer, and a 45 Mbps optical-electrical converter. Fujitsu claimed that its U135 could carry 3 DS-3 signals repeaterless distances up to 40 km. The converter could change an electrical DS-3 signal into an optical signal.

At about the same time as the Fujitsu announcement, Northern Telecom began marketing a fiber optic terminal designed for

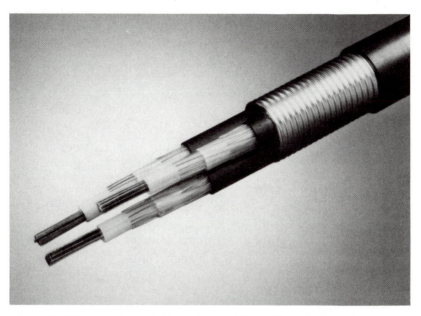

Figure 3. The number of fibers housed in cables continues to grow. AT&T upped the stakes considerably when it announced that it had available 408-fiber cable. (Photo courtesy AT&T.)

use on customer premises. The terminal could be used to link a business's telephone system or private branch exchange to the public telephone network, or as an interface between campus or building telephone systems and local telephone company central office switching equipment.

Such powerhouses as Du Pont and British Telecom have joined forces ostensibly to manufacture components geared to the coming local exchange market. Another Anglo-American alliance hoping to do the same thing is the U.K.-based Standard Telephones and Cables and Telco Systems, Inc. Still another example of United Kingdom interest in the market involved Pilkington's purchase of Phalo Optical Systems Division for approximately $5 million.

Another example of the acceptance of fiber optics in the local exchange marketplace—and its ability to be marketed there as a service—was when the Bell Operating Companies began offering tariffed service over fiber. Illinois Bell was first, followed by NYNEX, with more expected to follow.

Illinois Bell also announced the first fiber optic hubbing service, with the idea of giving long-haul carriers, interexchange companies and large end users easy access to one another's fiber optic networks. Early customers included US Telecom, WilTel, and Virtual Network Services. Using the Fiber Hub service, the long distance carriers are linked to an Illinois Bell hub—a termination point for DS-1 and DS-3 circuits at a central office in downtown Chicago. At the hub, Illinois Bell meets customers on an electrical or optical basis.

Illinois Bell also announced plans to triple its Novalink fiber optic network in downtown Chicago. Illinois Bell decided to spend an additional $1.1 million to install an additional 383 miles of single-mode fiber cable.

A compelling reason for the Bell Operating Companies to install fiber is to head off the threat of alternative carriers coming into their areas. Illinois Bell, for example, must contend with the Chicago Fiber Optics Corporation, which is innovatively installing fiber in Chicago's abandoned coal tunnels, linking major skyscrapers together.

One way to speed fiber into the home is for cable television and telephone companies to join forces, something that is easier said than done. For one thing, the FCC has prohibited telcos from the entertainment market, although there were indications that the FCC might be more lenient in allowing such matchings. Ohio Bell, for example, was selected by North Coast Cable Ltd. to install a fiber optic cable television system in Cleveland. As mentioned, Southern Bell is working its fiber-to-the-home experiment through Genstar.

Times Fiber Communications is coming back to fiber—now older and wiser—in hopes that this time the market will be ready for its cable television oriented products. In June 1986, American Lightwave Systems was formed, equally owned by

Times Fiber parent LPL investment group and by key management personnel who left Times Fiber to form the new company.

The Madison Lightwave Cable team mentioned at the beginning of this book was installing cable for bypass provider Institutional Communications Company (ICC), whose goal is to provide bypass throughout the greater Washington, D.C., area. ICC is using right-of-way provided by Bell Atlantic and three area utilities—Pepco, Vepco, and Baltimore Gas & Electric.

The leading alternative carrier as of the mid-1980s was Teleport Communications, Inc., which has linked a 150-mile regional fiber optic network with satellite earth stations on Staten Island. With the financial backing of Merrill-Lynch, Telport Communications is providing service for just about all the long-distance carriers that need to come into New York City.

A major boost to alternative carriers has come from the realization by corporate America that businesses can do better by designing their own communications networks—or using carriers who will do it for them—at a fraction of the cost of continuing to use conventional means. Stories of unresponsive entrenched carriers have helped to fuel the fire.

Competition has played a major role in the explosion of rewiring in both the cities and countryside, a competition that would never have come about without divestiture. But competition only provided the venue for this remarkable technology to demonstrate what it could do.

While the long-distance market had kicked off the rewiring of America, it has become clear that even greater amounts of fiber—and money to be made by using fiber—will come from rewiring the cities.

While various factors that we have explored determine how rapidly fiber optics came into both the cities and the countryside, the main factor of agreement has been that it will be the next generation communications medium of choice. As such, fiber optics will play an important role in this country's business dealings and will impact its overall growth.

Chapter 9

The Business of Fiber Optics

INSTITUTIONAL COMMUNICATIONS COMPANY (ICC), for whom Madison Lightwave Cable was installing cable in this book's opening scenario, was begun by Art Barber, of Bethesda, Maryland. Barber filed the necessary applications with the FCC and made some preliminary arrangements with right-of-way providers in the Baltimore-Washington area. Barber's office was in his house.

Nothing was heard from ICC for some time after the rudimentary papers had been filed with the FCC. One began to get the impression that this would be a company in name only. Then independent financier Scott Brodey took over the company with some heavy financial backing—to the tune of $20 million. Following the transaction, Barber would not deny the inside scuttlebutt that the deal had made him a millionaire. He acknowledged that he was "an entrepreneur" and really had no plans to see the venture through to its fruition in the first place.

Canadian Brian Hughes was having lunch in Washington, D.C., with Jonathan Miller, then managing editor of *Satellite*

Week, in the summer of 1983. The two were discussing the likelihood of success for Orion Satellite Corporation's bid for a private trans-Atlantic satellite system, with Miller noting that a private cable venture would not require Intelsat coordination and therefore had a better chance of success. Hughes, who had graduated from college six years before and gone on to serve as executive vice president of the satellite insurance brokerage subsidiary of Corroon and Black, was able to model a cable link well enough to build a business case for such a system. From the lunch came the formation of a new venture whose intention it was to build the first private trans-Atlantic fiber optic network. As we will learn in Chapter 12, this idea eventually boasted Cable & Wireless as a partner and received FCC approval. Miller abandoned the project by November 1983, but Hughes held onto his interest in the U.S. partner, Tel-Optik.

Big-time U.S. investors such as E. F. Hutton brought financing into Tel-Optik, diluting Hughes's share, but he stubbornly clung to 14% of the company. Regional Holding Company NYNEX has announced its intention to buy Tel-Optik for $10 million, if it can obtain a waiver from Judge Harold Greene regarding the modified final judgment and FCC clearance. If the purchase is made, Hughes will clear $1.4 million.

These two examples demonstrate the business opportunities that have occurred as the result of the coming of fiber optics.

The American business community knows about fiber optics, as it must know about any new tool that can revolutionize the way it does business. IBM, General Motors, Du Pont, Xerox Corporation, Honeywell, Wang Laboratories, Eastman-Kodak, Chevron, Fluor, Alcoa, Kaiser, Comsat—these are just a few of the major companies that have found out about fiber. Growing segments of the railroads, the utilities, and the financial community have found ways to deploy fiber optics.

A 1985 tour of AT&T's optical fiber manufacturing facilities in Atlanta, for example, found IBM officials trying to determine what type of optical fiber they wanted to better interconnect their mainframe computers.[1]

General Motors is developing ways to incorporate fiber optics into its automobiles and into the work place. GM subsidiary EDS is seeking ways in which GM can communicate better using fiber.

Du Pont's Connector Systems Group manufactures fiber optic components and a plastic strength member to be inserted in optical fiber cable. Du Pont has also used fiber optics to connect various facilities it has in Delaware so that the company can better take care of its internal business.

Du Pont has formed a joint venture with British Telecom, BT&D Technologies, to develop, manufacture, and market fiber optic components and devices. An early product is optical fiber equipment that allows a company to send signals in both directions along one fiber. Du Pont and Japanese subsidiary Du Pont Japan are developing fiber optic board level printed circuit data links.

Du Pont's involvement with fiber optics is interesting in that it represents a variety of ways in which a business can become enmeshed in fiber optics. For one thing, Du Pont has developed its own technology, plastic fibers, to compete in the fiber optics marketplace. This has, of course, been essential for U.S. companies to do—because the government was not there to provide the helping hand that was given by the governments of some other countries. With a few notable exceptions, the successful U.S. companies have been those large enough to have the deep pockets to invest in fiber optics, knowing that any payoff would be well into the future.

As has been pointed out, it is no coincidence that AT&T, with Bell Laboratories behind it, played a primary role in the birth and development of fiber optics. Corning, founded in 1851, has also

---

[1]"Special Report: AT&T Atlanta Works Tour Reveals IBM Presence," "Fiber Optics News," December 16, 1985.

had a rich tradition of research and development. As has been mentioned, Battelle Columbus Laboratories has offered up its facilities so that U.S. companies can try to design next-generation fiber optic components. One aim is to prepare for the more advanced Japanese products that continue coming into the U.S. marketplace. Early Battelle team members included Hewlett-Packard, Allied Corporation, Amp, Inc., Litton, and ITT. (See Figure 1.)

The options for small companies hoping to enter the fray on their own are limited. One possibility is to come into a promising market related to the mainstream of products but that is slightly different in scope—usually what is considered a niche market.

Figure 1. Robert Holman was the guiding force behind Battelle Columbus Lab's efforts to forge a corporate research drive into fiber optics. Holman has since left Battelle to work at Amphenol Fiber Optic Products (photo courtesy Amphenol Fibert Optic Products).

Gaining entry into a niche market must be accomplished early enough to capture much of that market and grow with it. This generally involves demonstrating enough discipline to put the necessary money into research and development on a regular basis to ensure that the technology is still the best and cheapest to serve that market.

One example of this is Galileo Electro-Optics Corporation, which was incorporated in 1973 as the successor to the electro-optics division of Bendix Corporation. From the time of its formation, which was early in the fiber game, Galileo has engaged itself in developing, manufacturing, and marketing fiber optic and electro-optic components that transmit, sense, or intensify light or images. Similar to Corning, Galileo established patents early in key fiber optic areas. While Corning protected its optical fiber manufacturing processes, Galileo protected fused fiber optic products, which are deployed in night vision components for military and security uses.

While the U.S. government has not provided large R&D monies, it has helped some companies by being a ready customer. This was, and still is, the case with Galileo. In its 1985 annual report, Galileo admits to a "long history of close cooperation with our national security agencies. . . ."[2]

Galileo's fiber optic products through the 1980s have consistently accounted for approximately two-thirds of its corporate sales. The company breaks its fiber optic products into two categories—fused and flexible.

Fused products are formed by fusing large numbers of optical fibers together to form a rigid structure containing thousands of fibers per square inch. The products can transfer an image on one surface to the opposite surface with minimal loss of resolution. Fused fiber optic products can magnify, minify, and invert images, and can save space and reduce optical losses when used instead of lenses and mirrors.

The company's flexible fiber optic products transmit signals or

---

[2]Galileo Electro-Optics, annual report, 1985.

images that employ much longer fibers than do the fused products. The fiber is used to supply remote illumination in flexible probes, display panels, and optical sensors. Products can also be used for viewing remote events and to inspect otherwise inaccessible or hazardous areas.

Galileo is committed to the importance of research and development. In its fiscal year 1984, for example, the company chalked up net sales of $19.2 million, had net income of $2.1 million, and put approximately $1 million into R&D. Galileo upped that R&D figure to $1.2 million in fiscal 1985. Net sales had reached $26.2 million by fiscal 1986. Net income soared to $3 million.

The Sturbridge, Massachusetts company has spent a considerable amount of money developing fibers that can operate in the infrared. It has developed fibers that are alternatives to silica fibers, including those made of fluorine and chalcogenide glasses.

Three Galileo employees-Charles DeLuca, Raymond Jaeger, and Mohd Aslami-left the company to form SpecTran Corporation, which is an optical fiber manufacturer that has registered some success in entering specialty markets. Peter Schultz left Corning to come to SpecTran, which is also based in Sturbridge. [Since then, Aslami and DeLuca have gone on to form another company, Automated Light Technologies, and Schultz has moved over to Galileo.]

Galileo went public at the beginning of 1983, offering 700,000 common shares. Galileo had a separate offering in 1986 of 500,000 shares. Shares of stock in the first offering ranged from $8 to $10; in the second, they were priced at $31 per share.

Galileo is now a company that includes more than 300 employees and is attempting to place more emphasis on marketing efforts. Not surprisingly, it hired a 14-year veteran of Corning, Josef Rokus, to be its vice president of manufacturing. In fiscal 1984, Galileo was still dependent to some extent on large customers (Xerox accounted for 20% of its business, Varo, Inc., made up 12%).

The company is also increasing expenditures in engineering costs. These costs rose in fiscal 1986 by 16% over what they had been the year earlier. Helping to account for the additional expenditures was a new line of fiber optic industrial sensors. The company also introduced a fiber optic bar code reader in early 1986.

Galileo has had its share of ups and downs, as have most businesses. (It recorded a loss of $177,000 in fiscal 1985.) But Galileo says it is "committed to developing the most advanced electro-optic and fiber optic products known to mankind. In fact, we're at the cutting edge of a new frontier that will significantly impact the lives of each and every one of us."

The story of Dorran Photonics is a tribute to the fiber optics industry. Dorran is that rare example of a company that came into existence to fill a mainstream void, without the need for extensive capitalization to get it going.

Spurred by the MCI single-mode fiber orders, the market almost overnight had shifted away from multi-mode and to this hot new commmodity. There was also a corresponding demand for single-mode fiber connectors. It was here that Dorran came in. The New Jersey based company could provide the product, so it got the business. As the result, in one year's time, the company went from zero employees to 120 employees.

This is not the gradual linear extrapolation that company managers enjoy plotting in their annual budget books, and the real fear that Dorran would decline as rapidly as it sprouted has been expressed on more than one occasion. To try to stabilize, the company attempted to diversify its product line to include multi-mode connectors and accessories. After climbing over the 200-employee level, Dorran did announce an 18% staff reduction in spring 1986 but does not anticipate further reductions.

Dorran has since been acquired by 3M Company, which would say only that it paid more than $10 million in a cash deal. Prior to the acquisition, 3M had attempted its own entry into the fiber optics market with an arc fusion splicer and cleaver. Purchase of Dorran—which 3M labeled the world's largest independent manufacturer of fiber optic connectors and other interconnection products—signaled that 3M meant business.

Obtaining funds from venture capitalists is yet another way a small company can stay afloat, giving it time to develop the R&D that will keep it alive. One example is Virginia based Fibercom Inc., which is attempting to come to market with high-speed fiber optic local area networks. Fibercom is backed by TA Associates, which claims it makes minimum investments of $2 million and is actively looking for fiber optic startups to invest in.

One group of investors that is receptive to funding small fiber optic companies is larger fiber optic organizations. Corning, for example, as of early 1987 owned 46% in the budding Raycom Systems, the Colorado based designer and manufacturer of fiber optic local data communications products. Du Pont, by the way, has signed a marketing agreement with Raycom Systems. The Du Pont deal represents another way large companies can gain influence in smaller companies, while the smaller companies enjoy the larger company's clout.

Another example of a larger company investing in smaller fiber optic companies is the purchase of shares in Lytel, Inc. and FOCS by connector manufacturer Amp. Inc. Amp has also established a technical and marketing affiliation with Lytel.

These affiliations are symbolic of the consolidation that has marked the fiber optics industry. Kessler Marketing Intelligence has estimated that the number of fiber optic companies doubled in one year—from 1984 to 1985. With that many companies, there has to be extensive overlapping of services.

In some instances, corporate mergers have impacted fiber optic operations. The merger of Allied and Bendix, for example, meant the consolidation of what had been two sometimes competing fiber optic connector manufacturers. By the end of 1986, Allied was looking for a buyer for subsidiary Amphenol Fiber Optic Products and six other companies.

In other instances, such as Corning's purchase of equity in Raycom Systems, companies are investing in both their own corporate futures and also for future business requirements.

Westinghouse became a minority equity partner in Infrared Fiber Systems, a company formed in 1986 to supply infrared transmitting optical fibers, fiber bundles, and bulk optical components for industrial, medical, and aerospace applications. The internationally recognized Celanese purchased a 10% share in Codenoll Technology Corporation, a manufacturer of fiber optic data communications products.

Corning is not afraid to get into joint ventures whenever it sees fit and this has become an operating part of company philosophy for many years. While Corning has received a straight patent license in Japan, it has developed successful joint ventures in Italy, France, the United Kingdom, Spain, and West Germany.

The Corning deal in West Germany was carried out with Siemens and became known as Siecor GMBH. Siemens came up with the so-called loose-tube design for cabling optical fiber, a design that Siecor chairman Allen Dawson believes "enabled us to literally switch overnight to single-mode."

Siecor in the United States was begun in 1977 in an old Robert Hall store in Horseheads, New York, with 14 people. In 1980, Siecor purchased a copper cable company but later sold the copper cable operation to another company. Siecor purchased a cabling facility in Hickory, North Carolina, and brought it into production in 1981. Capacity was quadrupled by 1984 and doubled again by the following year, according to Dawson. Siecor, in turn, created Siecor FiberLAN, whose main goal was to target the fiber optic local area network market. BellSouth has since bought a 50% share in FiberLAN.

But suppliers are not the only ones who have found out about the advantages of working together. Carriers trying to compete against AT&T are also finding it easier to do so by joining forces rather than trying to kill one another off. US Telecom and GTE Sprint have combined their networks to form US Sprint, an organization whose operations are heavily dependent on public acceptance of the nationwide fiber optic network they are building. It is interesting to note that Sprint and US Telecom had originally worked out a fiber optic capacity swap before joining forces, a move that helped to pave the way for the merger.

MCI's purchase of Satellite Business Systems from IBM is another example of network consolidation. LDX has become part of WilTel; WilTel has agreed to interface with Lightnet; SouthernNet and Southland are merging; and Allnet and Lexitel have agreed to join forces.

Small companies attempting to go public at an early stage in order to procure funding may run into trouble fast. That is because Wall Street often seems addicted to the syndrome of "instant gratification," hoping to turn a quick buck, according to Gordon Lamb, a leading investor in the area, who heads the Laser & Advanced Technology Fund. The years of R&D funding it often takes to bring a successful fiber optic product to market can make a poor match with investors seeking quick profits.

Du Pont's efforts to match its own technological expertise with the demands of the fiber optic marketplace are again in line with what other large corporations are doing. In Du Pont's case, the chemical giant is bringing its refined uses of plastics to fit a need in optical fiber cable—using Kevlar as a strength member in that cable. Kevlar provides strength, without adding a great deal of weight, helping to allow users to take advantage of one of fiber's basic qualities. Kevlar also allows the cable to remain dielectric, immune from any electrical signal, something that may not be said of cable that employs a steel strength member.

Kodak formed a division to market an expanded beam fiber optic connector, based in part on its vast knowledge of lenses. The connector uses precision aspheric glass lenses made by Kodak.

Dow Chemical is using its expertise to design a line of specialty optical fibers.

Yet another example of matching a company's expertise to satisfy marketplace requirements is the joining together of aluminum manufacturers—such as Reynolds Metals and Alcoa Aluminum—with cablers such as Siecor Corporation and Fujikura Ltd., to manufacture optical power grounding wire (OPGW) for use by the utilities. The wire has the dual advantages of grounding high power electrical lines and carrying communications.

Du Pont hired the Diamond State Telephone Company in 1985 to install a 30-mile fiber optic network to help Du Pont handle its high data and voice requirements. Du Pont paid out more than $15 million to interconnect 7 major nodes and 35 locations in New Castle County, Delaware. The network connects mainframes, mid-sized computers, and personal computers.

This represents yet another way in which business is getting involved with fiber optics—to satisfy its own communications requirements. In the case of GM and subsidiary EDS, for example, they were attempting to get around an annual telephone bill of about $1 billion. "If they can build their own nationwide network for $2 billion, it'll pay for itself in two years," observed NTN's Martin McDermott.

The financial community, encompassing both banks and brokerage houses, is also fitting the advantages fiber optics offers to satisfy its own communications requirements. Philadelphia's largest banking institution, Meritor Financial Group subsidiary PSFS, has decided to use fiber. Large portions of Wall Street are wired with fiber, as are investment communities in other cities. AT&T officials consider the wiring of Miami's financial district in 1984 as the first example of bringing fiber to the office.

In some instances, these financial institutions have had to shake off bad experiences with satellite providers before being able to invest in fiber. The advantages of fiber, however, have restored lost faith.

"The satellite industry in the last five years has come about much differently than was planned," explained Bank of America's Anne D'Andrade, speaking at a fiber optics conference in 1986. "Projections from the satellite industry are never met—at best!" she said. Because of problems the satellite industry encountered, financial institutions such as Bank of America "became much more partners with the satellite companies than we had planned."

Merrill-Lynch found out about fiber optics early in the game. New York Telephone, as part of AT&T, provided some early fiber optic subscriber routes, cutting over an initial link in February 1982. That joined Merrill Lynch's headquarters in Manhattan with its data center.

It soon became apparent, however, that Merrill-Lynch had more ambitious plans than to simply let the local phone company build its fiber optic network. Following completion of the New York Telephone job, Merrill-Lynch officials began talking about a new bypass entity, which was to become Teleport Communications.

It was obvious that Merrill-Lynch had learned much from its early work with New York Telephone/AT&T. The Teleport hired Robert Annunziata, who eventually was to become president; Annunziata had been an AT&T national accounts manager for the Merrill-Lynch job. Teleport also hired Howard Bruhnke to be its vice president of engineering. Prior to coming to Teleport, Bruhnke had spent 36 years working for New York Telephone, the final 8 years working with fiber optics. Bruhnke had also worked on the Merrill-Lynch installation. (See Figure 2.)

While known as a teleport, which connotes primary dependence on satellite earth stations for communications, Teleport officials readily admitted that it was the 150-mile regional fiber optic network that got the venture off the ground. Annunziata noted, for example, that the teleport's terrestrial fiber optic network was carrying over 1 Gbps worth of traffic in the first nine months of 1985. Teleport appeared destined to be the first major successful bypass carrier and has done such revolutionary things as signing up AT&T as a customer. Teleport also signed a fiber swapping agreement with New Jersey Bell, one of the telcos it is bypassing!

From its inception, the venture was firmly committed to fiber optics for providing its terrestrial requirements. With regard to microwave, for example, Bruhnke has noted: "Microwave will turn your hair grey; I guess I'm a living example of that. We don't believe in microwave."

Teleport Communications has also been firmly committed to

Figure 2. Teleport Communications President Robert Annunziata speaking at the teleport dedication on September 11, 1986. Annunziata had worked with AT&T as Merrill-Lynch's national accounts manager prior to joining Teleport. (Photo courtesy Teleport Communications.)

state-of-the-art technology and has not scrimped on dollars to get the best. (Annunziata acknowledged the venture had been "capital intensive." It is doubtful that—without Merrill Lynch's deep pockets behind it—the venture could have continued to foot expenses. The first year of profitability is expected to be 1988.)

Teleport, for example, was one of the first to use Telco Systems' M560 fiber optic electronics. The network employs riser cable to scale the 110 floors of the World Trade Center. In all, the teleport has from 25 to 30 contractors, including Artel Communications Corporation, which provided analog fiber optic equipment for major network feeds into and out of Manhattan. Siecor Corporation has provided cable and Western Union has done installation. Other suppliers have included Codenoll Technology Corporation. and the Grass Valley Group.

Teleport's own engineers claim they have when necessary invented support products. These products include inner ducting for underground conduit and an environmentally safe aluminum cladding for optical fiber cable to replace polyvinyl chloride, which is toxic when burned.

Teleport also used Telco Systems' Teltrac control monitoring system, which allowed the teleport's operators to locate and correct immediately network flare-ups, sometimes within a minute's time. "Teltrac allows me to run the system," explained Bruhnke. "In the past, the system has run me." Another major customer of Teltrac is Sears.

The list of customers has grown considerably as the venture has become known. Merrill-Lynch became a major customer, as did Dow Jones. Other customers include Bankers Trust, the Securities Industry Automation Corporation, IDB Communications Group, Ltd., and the American Satellite Company.

Teleport's success can be clearly linked to the increase in private networks. Teleport can often provide service, for example, to companies more rapidly than can New York Telephone and in ways that are more responsive to the customer's requirements.

In March 1987, Teleport appealed to the FCC to allow it to connect New York Telephone's offices with its own long-distance customers. Teleport promised to pay for the privilege but faces stiff opposition from New York Telephone.

The effect of businesses designing their own private networks is that they have enlivened the marketplace. While AT&T and the Regional Holding Companies provide stiff competition, these options have allowed others to gain a foothold. Sometimes the company playing the trump card that gets the job is the carrier that can provide fiber optic capabilities in a reasonable period of time.

Du Pont's joint venture with British Telecom, BT&D Technologies, is an example of how British and other European concerns are trying to make inroads into the lucrative U.S. marketplace.

Other British companies have come into the U.S. market utilizing a variety of avenues. Cable & Wireless joined with the

Missouri-Kansas-Texas railroad to form Electra. Pilkington bought out the California division of Phalo Corporation's Optical Systems Division. Plessey Ltd. purchased a division of Stromberg-Carlson and has landed contracts with what is now US Sprint and with Northwestern Bell—the former for high-speed electronics, the latter for video transmission capabilities over fiber. Plessey had also joined with Corning to form fiber optics device manufacturer PlessCor, but subsequently withdrew. The corporate name was changed to PCO, Inc., to reflect the change.

Standard Telephones and Cables has joined with Telco Systems, Inc., to offer fiber optic transmission equipment geared to the U.S. fiber optic local loop marketplace. The first two products are 6.3 Mbps and 140/280 Mbps systems geared for that marketplace. STC also landed a contract to provide optical fiber cable to what was then US Telecom.

Both Telco Systems and STC provided equipment for the first products with systems integration occurring at Telco Systems Fiber Optics Corporation. The products carry the Telco Systems label and are marketed in the United States by Telco, which provides customer training, equipment installation and service, and field support.

For its part, Telco Systems has access to Standard Telecommunications Laboratories of the United Kingdom, which served as the locale for the historic Kao/Hockham paper. STC has developed "incredible drive technology" since then, according to Bowman.

Founded in September 1972, Telco Systems had been a relatively small California based concern specializing in voice frequency products.

This changed drastically when the Raytheon division was bought. The resulting subsidiary, based in Norwood, Massachusetts, was to become Telco Systems Fiber Optics Corporation, to be presided over by Bowman, who came from Raytheon. The agreement was consummated on January 13, 1983.

In its 1986 annual report, Telco Systems advertises itself as a company that "designs, manufactures and markets digital transmission equipment and systems for the telecommunications industry.[3] This represents a departure for the company, which in the past said it was a "manufacturer of digital and analog transmission equipment."[4]

The advent of Telco Systems Fiber Optics Corporation, however, became the center of attention and exhilarated the U.S. fiber optics marketplace because the company in 1984 said it had available electronics operating at 560 Mbps, its so-called M560 product line. This was in addition to a host of lower transmission rate products. [Other Telco Systems products include the 828F (45 Mbps), M90 (90 Mbps), Tel-Loop (which lowers installation costs), and Teltrac (the aforementioned network monitoring system).]

The prospect of a U.S. company coming in to provide electronics faster than the 400 Mbps equipment the Japanese were offering caused a major stir. With AT&T's failure in this area, the interest in the Telco Systems product line became intense.

While Telco Systems Fiber Optics Corporation had started out with approximately 40 employees, it soon became apparent that big things were happening. Only 18 months later, the company had ballooned to 250 people.

In fiscal 1984, the parent company's sales were approximately one-third fiber optics, one-third autodialers, and one-third voice frequency products, according to then chairman and CEO Philip Otto, who had been with the parent company prior to the Raytheon acquisition. For the quarter ending November 1984, the first fiscal 1985 quarter, that division which had dramatically shifted to 60% fiber optics, 25% voice frequency, and 15% autodialers and products that came from another acquisition, the Illinois based Telebit.

With fiber optics growing by leaps and bounds, Telco Systems

---

[3]Telco Systems, Inc., annual report 1986, "Positioning Ourselves for the Future."

[4]Telco Systems, Inc., annual report, 1985.

Telco Systems seemed an exciting place to be. The company announced it was going public in February 1984. By the beginning of 1985, it had procured a strong buy recommendation from respected Wall Street brokerage house Salomon Brothers. Fiber Optics Corporation announced major M560 contracts with Bell Atlantic, Pacific Bell, and a number of other carriers. The company had hundreds of orders for the equipment.[5]

Then the roof caved in. In 1985, it became apparent that Telco Systems was having major problems with its M560 electronics. The equipment was due to arrive by the beginning of the year, but Bowman acknowledged there would be a six-month delay. He blamed the situation on "software and firmware problems and some of that required design adjustments to the hardware."

As the result, Telco Systems lost its contract with Pacific Bell. It was later required to pay Pacific Bell $960,000 because of the missed deadlines. In other instances, Telco was placed in the somewhat humiliating position of having to subcontract NEC America to provide NEC's 400 Mbps electronics until the Telco Systems equipment was ready. NEC had to come in and provide the electronics for the Bell Atlantic Washington Metropolitan Area Lightwave System, a job that Telco had earlier showcased as a good example of its new electronics.[6]

In addition to these expenses, which caused the company to establish a reserve of $1.5 million to cover costs incurred by the substitution of products and led to the company's doing some jobs for little gain or at a loss, Telco Systems had to write off its autodialer line at a loss of $7.5 million.

Even with the headaches, the company characterized development of its M560 Mbps fiber optic electronics as "the most significant achievement" in its corporate history. While noting that there was customer disappointment with the delay, Telco

---

[5] "Telco Systems Already Has 'Several Hundred Orders' for 560 Mbps System," *Fiber/Laser News,* June 8, 1984.

[6] "Telco Systems Had to Rely on NEC Equipment for First Phase of Washington Lightwave System," *Fiber Optics News,* September 30, 1985.

Systems still noted that it was "one of the first into the market-place" with the 560 Mbps electronics. (The first was provided by N. V. Philips in spring 1985 to AT&T.)

Telco Systems characterized 1985 as the year fiber optics played the predominant role, chalking up 79% of sales. The company's Voice Frequency Products Division, which is what was left of the West Coast operation following write-off of the autodialer unit, was combined with Telebit, which was purchased in August 1984 to form the subsidiary Network Access Corporation. (This meant the parent company had two subsidiaries, Telco Systems Fiber Optics Corporation and Network Access Division.)

Bowman was promoted to president of the parent company and the former head of ITT Electro-Optical Products Division, Jack Shirman, came in as president of Fiber Optics Corporation. Clearly, the company was being driven by its fiber optics products, and Otto's star was in decline. In mid-1986, the company announced that Bowman had replaced Otto as chairman and Chief Executive Officer, although Otto was to remain on the Board of Directors. Shirman was to become president of the parent company, as well as stay at the helm of Fiber Optics Corp. The company announced that it would shift its corporate offices to Norwood, Massachusetts, from California.

The company certainly had been suffering from growing pains, and this no doubt led to some of the turbulence. By August 31, 1985, it had 855 employees and recorded total sales of $94 million, $79 million of which came from fiber optic products.

Bowman said the company believed all along that the local exchange would be where its products were best suited and, if early results from Teleport were any indication, the local exchange market could provide the company's salvation, particularly with STC at its side.

But there were more problems to come. In fiscal 1986, corporate sales slipped to $89.7 million, and a net loss of $8 million was recorded. Bowman, like Otto before him, was slipped out of his position, with Shirman taking the helm. Bowman has since resurfaced as chairman and CEO of another, albeit smaller, fiber optics company, Artel Communications Corporation.

The beginning of Telco Systems' 1986 annual report spoke volumes: "1986 was the year we came face to face with reality," it said. "We realized that, like many a young company before us, we had grown too fast—1200% in three years—without building the management structure to control that growth."

"We recognized that our edge is in fiber optics, not voice products," it laments later on. As the result, the company's Lombard, Illinois, facility was closed and more than 100 employees were laid off.

Telco Systems is not the only company to see fiber optics play an increasingly important role. Fiber optic pioneers Corning and AT&T have both seen fiber optics play increasingly important roles in their corporate affairs. Corning promoted Dr. David Duke to head its research laboratories, marking the first time someone associated with fiber optics had taken over that slot.

A Corning corporate realigment in 1986 created a division for electronics and telecommunications, of which fiber optics played a central role; before those activities it had been somewhat buried in the capital goods components group. Corning also continued to update its product periodically, announcing dispersion-shifted fiber in 1985; 62.5 × 125 multi-mode fiber in 1986; and specialty fibers, including hermetic-coating fiber, in 1987.

In fairness, Corning may have overemphasized the role of optical fiber in its operations. The company had the capability in 1987 and 1988 to manufacture more optical fiber than is anticipated in the entire U.S. marketplace during those years, according to its own estimates. Corning estimates the market is receding, but predicts the decline will not be as severe as some other prognosticators, such as Kessler Marketing Intelligence.

As the result of the softening that occurred from the second half of 1986 on, Corning had to lay off 180 employees and reassign an undisclosed number of others. Corning's entire fiber optic manufacturing operations were shut down from Thanksgiving 1986 through New Year's Day 1987.[7]

AT&T, for its part, tried to develop more marketability for its fiber optic operations, particularly in the components arena. The company since divestiture tried to sell cables, connectors, splicers, and apparatus, in addition to the systems for which it was well known.

The West German based Siemens acquired a majority interest in GTE's transmission business, including the Transmission Products Division in the United States. This included transmission systems for trunk and loop applications, including digital fiber optic terminals, repeaters, and multiplexers.

The French Societe Anonyme de Telecommunications scored a contract with Litel Telecommunications to provide its 560 Mbps electronics. CIT Alcatel U.S. subsidiary Citcom Systems also had some success in bringing its 565 Mbps electronics to the U.S. market. SAT was also considering becoming an equity partner in Litel, but ownership assumed by the Italian company Telettra and Pirelli straddled the 25% foreign limit FCC regulations allow.

Via an arrangement with CGE, owned by the French government, ITT's optical fiber and cabling operations are now under French ownership. Included is the much-traveled Valtec name.

Ericsson also brought its high-speed electronics to the U.S. market although—like Telco Systems—could not come to market when promised and forfeited contracts as a result. Ericsson has also had success selling optical fiber in the United States.

NKT of Copenhagen, Denmark, also attempted to bring high-speed electronics to the U.S. market.

AT&T used NKT's cabling operation to try to establish a European beachhead. Near the end of 1986, AT&T purchased majority ownership in NKT's fiber manufacturing plant in Denmark, and immediately announced plans to triple production levels.

U.S. companies are trying to break into Far Eastern markets but are having considerably less success than is Japan in the U.S. marketplace. The United States imported more than $98 million in optical fiber and cable in 1985 and nearly $105 million in 1986, according to a Commerce Department study entitled

---

[7]*Fiber Optics News*, Jan. 26, 1987, p 1.

"1987 U.S. Industrial Outlook." U.S. exports, on the other hand, were almost $64 million in 1986, up from 1985 export figure of $34 million.

Corning already claims Fujikura and Furukawa as licensees in Japan, but otherwise has had limited success impacting the Japanese market. Corning did sign a joint R&D agreement with NTT to explore further the use of fluoride glasses, which are expected to improve the importance of optical fiber. Corning has in the past told NTT that it paid more for optical fiber cable than it should have. NTT did order optical fiber cable from Siecor at the end of 1986 and has also ordered optical fiber cable from AT&T.

U.S. suppliers are having more success in other Asian countries, however. ITT has executed a technology transfer with major South Korean concern Samsung. An optical fiber cabling facility there is to have the capability of producing 10,000 fiber km annually.

AT&T signed an agreement in 1984 with major Samsung competitor Lucky Goldstar Group to provide technology, training, and equipment to manufacture fiber optic systems in South Korea. AT&T also signed an agreement to help build an optical fiber cabling facility there.

In 1986, AT&T sent its first order of optical fiber cable to the People's Republic of China. The multi-mode fiber was to accommodate three fiber runs connected to a 5ESS switch, also the first to be sent to China. Corning hopes to establish an optical fiber manufacturing facility there and has held talks with the Chinese to that effect.

Yet a problem for U.S. suppliers in dealing overseas is that they are sometimes limited by Defense Department concerns, particularly when dealing with countries such as China. Some fiber optic products have been placed on a military critical technologies list and therefore cannot be sent to Communist countries. Unfortunately for suppliers, these are often the most technologically advanced components—the products these countries are often most interested in receiving.

Yet there are valid reasons for not rushing these products to whoever wants them. The Soviets, for example, used fiber optics extensively to bug the U.S. embassy in Moscow. The technology was taken from outside the U.S.S.R.

The gains made by Japanese companies Fujitsu and NEC coming into the U.S. electronics market have already been chronicled. Lens supplier Nippon Sheet Glass also continues to have a presence in the United States. Fujikura has entered via the optical power grounding wire venture with Alcoa.

Companies such as NEC, Fujitsu, and Sumitomo Electric have all built multimillion dollar facilities in the United States, confident that they will continue to score gains in the fiber optics markets. The Sumitomo landing is a little more risky, considering the threat of Corning legal action hanging over the venture's head.

Corning and Sumitomo have continued to slug it out in the legal boxing ring. When the International Trade Commission ruled that Sumitomo could not be kept out of the U.S. market because it was not monopolizing the marketplace, Corning filed suit in the U.S. District Court in New York. The ITC ruling noted that Sumitomo was violating Corning's patents. The Corning action requested that an injunction be slapped on Sumitomo, enjoining Sumitomo from future sales of the optical fiber in the United States, claiming that three of Corning's patents were violated. This suit—and the Sumitomo suit filed earlier in North Carolina by Sumitomo against Corning—were eventually brought under the same roof in New York. Sumitomo was not barred from supplying fiber and continued to command a small market share.

Sumitomo was later to sue Corning for patent infringement. The Japanese company unabashedly pushed its Z-fiber, which is fluorine rather than germanium doped, and which Sumitomo claimed had the "lowest achievable attenuation." (See Figure 3.) Sumitomo has also landed a contract with the US navy to provide optical fiber cable to tether undersea remotely-operated vehicles with manned control units.

Federal District Court Judge William Conner in October 1987 found Sumitomo in violation of two of Corning's patents and Corning will be awarded damages in the case. The two patents found to be violated included Corning's 915 patent, which cov-

Figure 3. Sumitomo has not been inhibited from the threat of Corning legal action to launch an aggressive advertising campaign promoting its Z-fiber, which it claims provides the "lowest achievable attenuation." Sumitomo, however, was found by a U.S. District Court judge in violation of two of Corning's patents. (Source: Sumitomo.)

ers optical fibers used in long distance transmission, and the 550 patent, which protects fiber used in short-distance transmissions. Corning Chairman James Houghton said the decision "confirms the significance of the contribution that Corning has made to this technology."

Another factor in the equation is that Corning's key fiber patents are set to expire in the late 1980s and early 1990s.

To perhaps further distance itself from its competitors, Corning introduced dispersion-shifted optical fiber in February 1985 at OFC '85. Corning claimed dispersion-shifted fiber could provide the lowest losses at 1550 nm wavelengths, what many believed was the optimal wavelength for low-loss silica fiber. Corning introduced the fiber in experimental quantities.

However odious patents may be to a company such as Sumitomo, Corning and AT&T have used them as an effective business tool that have helped keep their optical fiber businesses thriving.

When the previously discussed SpecTran Corporation joined forces with Southern New England Telephone Company to form Sonetran, it found out just how strong the alliance is between the two companies that developed fiber optics.

Sonetran was created to manufacture optical fiber cable. Because it was 51% owned by SNET, the venture claimed that it would not have to pay licensing fees to Corning, since it had access to the same AT&T patents that Corning did under the original AT&T-Corning agreement. SNET claimed that—since it had an affiliation with AT&T for approximately 100 years—it should have access to AT&T's patents as well.

The venture had the potential to split the original Corning-AT&T agreement and pit the two against one another. But AT&T was having none of it. Even if SNET and AT&T had had an affiliation for that long, AT&T was not about to turn its back on Corning. Sonetran has since cut back large portions of its work force and the venture appears to be in serious trouble. The corporate ties that had first forged a technology remained unbroken.

Chapter 10

Revitalizing the Computer Industry

THE IDEA OF transmitting data by using fiber optics has been around at least as long as the first OFC conference back in 1975 and was recognized by the pioneers as an excellent potential application. Because it uses binary code (through pulses) to transmit information, fiber optics is considered a good match with computers, which also communicate via ones and zeros.

Today, fiber optic data systems are as common to everyday life as the 7-11 Store, Burger King, or Pizza Hut restaurant, where optical fiber cable can link the cash register with a data storage unit on the premises. The data unit in turn is polled by an off-site central data computer using conventional transmission.

An early battleground between communications supergiants AT&T and IBM has been that of providing local area networks (LANs) to customers. LANs are an efficient, expeditious way for office workers to communicate within a defined network. Both superpowers offer optical fiber cable as one way of accomplishing this. (See Figure 1.)

Fiber optics is expected to play a role in the coming of integrated services digital networks (ISDN), whose intention is to

Figure 1. Bringing optical fiber cable into buildings has become commonplace. New buildings can advertise an optically fibered structure as being more advanced for communications than more conventional wiring schemes. Here Amp employees install fiber under carpet. (Photo courtesy of AMP Incorporated, Harrisburg, PA.)

provide users all over the world with the capability of simultaneously talking and transmitting data to the outside world from their work stations. The large capacity that fiber optics allows is considered a plus for voice-data integration.

In the future looms optical computing, a way to store incredibly large amounts of information and provide extraordinarily fast response times. As we will see, fiber optics will also play a role in this exciting new frontier.

Bell Laboratories' W. M. Muska presented a paper at the first OFC conference in 1975 entitled "Detector-Taps for Single Fiber

Data Bus,"[1] in which he proposed that "an optical fiber data bus could be an attractive substitute" for metallic data buses; that is, networks could be used to carry messages between a central location and several remote locations by using optical fiber.

Muska cited many of the advantages proponents of fiber optics for other applications had used. These included immunity from electromagnetic interference, elimination of cross talk, lighter weight, and smaller cross-sectional area. Fiber optics also offered electrical isolation between units.

In his paper, Muska explained that optical fiber using plastic cladding outside silica glass might be more workable than the silica fiber being used for voice communications. Tapping into the network frequently with different kinds of computer equipment meant that the fiber had to be malleable, something the plastic cladding could provide. The higher losses the plastic-clad silica fiber introduced were not of major consequence, Muska thought, since signals only had to travel hundreds of feet in these localized networks. The ease of handling of multi-mode fiber, compared to single-mode, made it the fiber of choice.

When one speaks of data communications, it is generally helpful to divide the applications into two categories: 1) data communications in a point-to-point arrangement, and 2) data communications in the more defined, more closely knit LANs. Both are transmission mediums and are therefore different from data storage, although optical storage represents an advancement in computer storage capabilities expected to go hand in hand with optical transmission in next-generation systems.

Point-to-point data applications operate similar to voice networks. In fact, the integration of voice and data over the same network is an important goal. The whole idea behind ISDN is to provide users with the same data capabilities as they now have for voice, and as readily.

---

[1]"Detector-Taps for Single Fiber Data Bus," by W. M. Muska, Bell Laboratories, presented at OFC '75.

Data communications increasingly is becoming a driving force behind which a network is configured in the first place. One example of this is the General Motors EDS network, which is primarily CAD-CAM (a data service) driven.

LANs are more closed networks that connect computer equipment on a local campus, such as a company's office workstations located on various floors of a building or with offices in buildings nearby. A LAN may connect various terminals, printers and other types of computer equipment to a central controller, with each workstation having access to the company's computer power. Examples of LANs are in-house telephone networks, a computer and peripherals, a computer and remote terminals, word processors connected for electronic mail, a cable TV system, or combinations of some or all of these.

As with many distinctions, this one can be and is easily blurred. Many so-called LANs, for example, are really more point-to-point systems. Some early fiber optic based LANs, according to John Kessler, were called LANs but in reality used a passive tap permitting them to get off the fiber and then back onto the data bus, thus not truly conforming to the LAN definition.[2]

A touted advantage of fiber optics is its ability to provide voice, data, and video services along the same network. Yet by the mid-1980s, there was little evidence this was in fact occurring. A survey done by the California based CTPS Inc., between April and July of 1986, for example, found that only one to two percent of surveyed companies had made "any real voice/data equipment installations." Sherry Geddes told the trade newspaper *Communications Week:* "We've been doing a great deal of survey work and have discovered that, contrary to industry perceptions, voice and data [integration] just isn't yet being accepted as a viable method of moving information."[3]

---

[2] "Exclusive Interview: KMI's Kessler Predicts $3 Billion Market by 1989," *Fiber/Laser News,* July 22, 1983, Phillips Publishing Inc., Potomac, Maryland.

[3] "Voice and Data: A Marriage Made in Heaven or Hell?" by Bob Johnson, *Communications Week,* Aug. 18, 1986, p. 9.

Yet the same publication also reported that user groups representing either voice or data concerns must "integrate their interests or—quite literally—die."[4] The convergence of voice and data will come about as optical fiber moves closer to the subscriber premises, according to Kessler.

This early integration of voice and data is but one of many indications of the importance communications providers are attaching to ISDN as the goal of the future. The foregone conclusion of these carriers is that ISDN will be a reality not only in America but throughout the world, a key ingredient in the Global Village concept.

By 1984, contributors to the Internationl Telegraph and Telephone Consultative Committee (CCITT)—the organization responsible for developing ISDN—had already suggested an optical fiber interface with the user, according to Anthony Rutkowski, who attended meetings as an FCC representative. "The term broadband ISDN was coined to describe such an interface compatibility," he noted. The term "was subsequently defined to always imply transmission channels capable of supporting rates greater than the primary rate, with examples extending up to 560 Mbps."

The intense interest in fiber optics in ISDN formulation has led to a situation, according to Rutkowski, where the question "is no longer whether fiber-based broadband ISDN implementation will overtake those based on twisted pair, but when."

Nippon Telegraph and Telephone (NTT) in 1985 proposed that fiber optics be a major transmission medium in the development of worldwide ISDN standards. NTT presented its paper in Kyoto, Japan, before government and industry communications representatives from countries throughout the world.[5]

AT&T's goal of having a largely digital network by 1990 is certainly driven in large part by the fact that it will have large

[4]"'Integrate or Die' Is Motto of Voice and Data Groups Since Divestiture," *Communications Week*, May 12, 1986.

[5]"NTT to Propose Fiber Interface for International ISDN Standards," *Fiber Optics News*, November 11, 1985.

amounts of fiber optics to use as part of that network. The National Telecommunications Network intends to make its network "ISDN compatible." Through its Telenet company, US Sprint is integrating data services over the company's nationwide fiber optic network.

Spearheading the ISDN trials in the United States have been the Bell Operating Companies. The telephone companies should remember that installing fiber optics prior to the time ISDN comes about can work in their favor when that does happen, according to Corning's Scott Esty. Commenting in 1985, Esty stated: "Companies can build huge 'digital pipes' now by installing optical fiber cables with today's transmission technology. When ISDN protocols are defined in the late 1980s or early 1990s, the in-place fiber optic 'digital pipes' can be upgraded by replacing the end electronics, probably with capacity to spare. By providing integrated voice, video, and data transmission, ISDN will provide transmission services that end users soon will demand."

Working with AT&T, Illinois Bell tested a network in the Chicago area. US West used a 141-mile fiber optic trunk in the Phoenix area to conduct several ISDN trials. In one, Northern Telecom was to employ its DMS-100 switch to provide ISDN services for multiple customers along a 28.8-mile optical fiber route. A second application involved using an AT&T switch over a 22-mile route. A third used a GTE switch to transmit simultaneous voice and data to two remote switches.

It is appropriate that carriers such as AT&T and the Bell Operating Companies should be at the forefront of ISDN design in the United States, since they were the ones that painstakingly brought voice service to America. One example of AT&T's commitment to bringing ISDN services over fiber involved the introduction in 1987 of the optically remote module (ORM), considered by AT&T to be the world's first remote switch module to connect to its host via a standard fiber optic system.

The first AT&T ORM was located in the Phoenix main central office as part of the ISDN test. The remote switch is connected to the AT&T 5ESS switch communications module, located 25 miles away in Chandler, Arizona. ORM can be used to extend ISDN "with minimal capital investment into analog switching offices, thereby creating digital overlay networks," according to AT&T.

Yet when it comes to providing localized data interconnect in an office environment, IBM has assumed the preeminent position. There is no getting around the fact that IBM has been slower to adapt the advantages of fiber optics than other companies, although it can be argued that IBM's need for fiber has been slower to evolve than have the requirements of the transmission carriers such as AT&T.

IBM's entry into the transmission business was via the money-losing Satellite Business System; this may have revealed an over-emphasis within IBM on satellite. This was corrected to some extent when IBM sold SBS to MCI for a sizable share in the more fiber-oriented MCI.

Still, there are indications that IBM intends to gravitate to fiber in its own way for its own products. IBM's Robert Armstrong told a panel at FOC/LAN '85 that there will "obviously be an evolution to fiber over time" in LANs.[6] IBM has already found several ways in which to integrate fiber optics into its corporate offerings and is working on other exciting possibilities.

One product is the IBM 3044 channel extender, which uses 62.5 × 125 micron multi-mode fiber to extend distances between workstations, display terminals, graphic terminals, and other processors. The channel extender can join as many as 33 peripherals to a mainframe computer in a LAN. Yet IBM was not the first to come out with such a channel extender, and some sources do not believe it offered the best. In fact, there are indications that IBM may have been prodded into developing the channel extender by competitors.

Connecticut based Data Switch, for example, preceded IBM to

---

[6] "Evolution to Fiber Seen By IBM Official," *Fiber Optics News,* September 23, 1985.

market with its Syslink fiber optic system, which can connect mainframes up to 3.4 miles apart. At the time of the product's introduction, former IBM official Robert Gilbertson, president of Data Switch subsidiary Channel Net, said his company could potentially increase the distance to 20 miles. The capacity that fiber optics provides could allow two IBM or IBM-compatible mainframes to "operate as one computer," according to Gilbertson.

One example of Data Switch's success was a $700,000 order from Ford Motor Company for channel extenders at data centers in Detroit. Data Switch in December 1986 was claiming that its channel extenders could enable IBM and IBM-compatible computers to communicate with high-speed T-3 circuits and with tape drives.

One result of using channel extenders is that the clogged up computer rooms that we have come to know will no longer exist, since various computer components can be linked at different places throughout the network.

A second fiber optic type application developed by IBM and marketed through its Rolm subsidiary is interconnecting PBX nodes with fiber optics. Here again the company uses 62.5 $\times$ 125 micron multi-mode fiber.

A third way IBM incorporates fiber optics is by offering an optical fiber wiring scheme for its token ring local area network. Yet it must be remembered that the type five wiring scheme is only an option to what admittedly has evolved as a copper based network. Unlike the other two offerings, IBM uses 100 $\times$ 140 multi-mode fiber for this. IBM has also developed the 8219 optical fiber repeater to extend to 2 km the distances between wiring closets and buildings.

Apple has introduced a fiber optic local area network designed by Du Pont Company for Apple's Macintosh. The scheme is a refined version of the "Appletalk" network. '"Appletalk" networks can now readily connect three times the usual devices and go five times the regular distance between stations, with a total network length that is practically unlimited," according to Jim

Somerville, who heads Du Pont's opto-electronics systems marketing effort. More than 100 "Macs" can be interconnected.

Du Pont and Apple are gunning for both the commercial and military/government market sectors. Early customers include Armco, Inc., a New Jersey based Fortune 100 company, and Systematics General Corporation in Virginia, which is developing a Tempest-approved fiber cable network system for data security specifications.

That IBM has assigned only minor importance to optical fiber cable says something about LANs, how fiber optics fit into them, and IBM. To say that the use of optical fiber in LANs has evolved slowly is a gross understatement.

Companies have taken major losses because the market has not evolved as rapidly as some had thought. A fundamental reason for this is that optical fiber cable has not lent itself to being continually spliced and connected to service the close-in requirements of LANs.

Copper, on the other hand, has been perfectly suited for handling the lower speeds and numerous ports required to connect computer equipment. The standard for LANs that evolved, Ethernet, was based on copper, with a standard speed of 10 Mbps, but with the capability of being upgraded to 50 Mbps. This worked well in satisfying IBM's basic requirements.

Companies such as IBM, Wang, and Burroughs relegated deployment of optical fiber cable for specialized applications, such as where security and immunity from EMI were key considerations.

Still, there is tacit agreement within the industry that customers will want LANs that have greater capabilities, that can employ higher transmission rates, and that could eventually replace Ethernet standards. This is where fiber optics is expected to take over.

The American National Standards Institute (ANSI) has laboriously undertaken the promulgation of standards for fiber optics to be used in these higher grade LANs through its X3T9.5 technical committee. The Fiber Distributed Data Interface (FDDI) standards, at least at the early stages, were based on transmission rates of 100 Mbps, ten times the Ethernet standard. The standard calls for a ring network, similar to IBM's token ring, consisting of stations logically connected as a serial string of stations and cabled fibers to form a closed loop. While every effort is being made to interconnect these LANs with existing Ethernets, the eventual goal is to replace Ethernet LANs, which are metallic based, with the much faster, less noisy, more secure fiber optic LANs.

A thorny issue the ANSI committee has tried to resolve is what kind of fiber should be used. It should be noted from the outset that multi-mode fiber was the dominant fiber of choice through the mid-1980s for LAN type applications and it is expected to remain that way at least until 1990. The easier handling and less demanding transmission rates have placed it in that role. While work continues, the committee seems to favor the 62.5 $\times$ 125 fiber, although there is room for alternatives, such as 100 $\times$ 140 micron core fiber. Trellis Communications Corporation President Richard Cerny believes that AT&T and IBM will agree that 62.5 $\times$ 125 fiber will be the way to go.

As an example of problems that a lack of standardization has caused in the past, a General Electric campus wired for the Defense Department was to include computers manufactured by both Wang and Burroughs. Wang insisted on 50 $\times$ 125 micron fiber, Burroughs wanted only 100 $\times$ 140 micron fiber. As the result, the two computer systems could not talk to each other, according to sources involved with the program.[7]

The FDDI standard was still evolving as 1987 came to a close. (There is already a second, FDDI-2, under development.) While the first physical layers of the standard were ready by the beginning of 1987, matters such as station management and types of connectors to be used still required resolution.

Once FDDI does come about, companies with representatives

[7]"Major GE Defense Campus Wired With Fiber," *Fiber Optics News*, January 13, 1986.

on the committee will form a nucleus for the industry, according to Larry Green, director of advanced technology at Advanced Computer Communications, and an X3T9.5 member. Green categorized the members as follows:

Mainframe companies: IBM, DEC, Sperry, CDC, Honeywell, and ICL

Peripheral companies: STC, Magnetic Peripherals, and CDC

Systems houses: ACC, NCR, Fibronics, Northrop, Martin Marietta, Hughes

Workstation companies: Apollo, Kodak

Component manufacturers: AT&T, Siecor, PCO, AMP, and Lytel

VLSI manufacturers: AMD, National Semiconductor

Users: NASA, NBS, U.S. Navy

The fact that FDDI is not in place has not inhibited companies such as Artel and Proteon from introducing FDDI type LANs. By the end of 1986, Artel had already shipped prototype units to the State University of New York, McDonnell Douglas, and Sandia National Laboratories.

AT&T, the Regional Holding Companies (RHCs), and independent suppliers recognize the importance of positioning for this emerging new market sector. Of the $1 billion market for LANs estimated in the United States by 1990, a healthy share is expected to go for fiber optic based LANs, and that percentage is expected to grow. ERA Technology Ltd., based in the United Kingdom, estimates that 7500 fiber optic based LANs were installed worldwide in 1986.

IBM is already experimenting with a computer chip that could have future applications for FDDI type systems. The chip can "read" photonic signals and "translate" them into computer language, according to IBM. In experimental tests the chip already operated more than twice as fast as any similarly developed chip set.

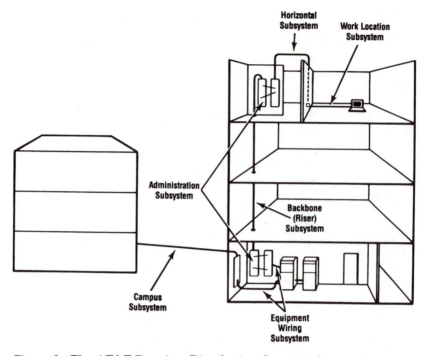

Figure 2. The AT&T Premises Distribution System relies on fiber optic cables to large extent. (Graphic courtesy AT&T.)

An AT&T-IBM square-off seems inevitable, as AT&T pits its better understanding of point-to-point networks and fiber optics against IBM's inherent familiarity with LANs.

AT&T's premises distribution wiring scheme, unlike IBM's token ring LAN, is primarily fiber optic based. (See Figure 2.) AT&T intends to have what it characterizes as the 100 top data markets in the United States wired for fiber as part of its surge through 1989. By the end of 1988, customers in selected areas will be able to transmit data via an uninterrupted fiber optic system joining the Far East and Europe.

Fiber and other digital media will provide AT&T with the capability of having advanced digital communications systems linking 350 major cities in the United States with other locations around the world. AT&T expects to spend approximately $2.5 billion on digital network expansion in 1987. Approximately 50% of the company's traffic will be digital by the turn of the decade, according to AT&T spokesman Rick Brayall.

Information will be transmitted more clearly and sharply because the system is digitally based, according to Frank Blount, president of AT&T's Network Operations Group. Calls will go through on the first try, due to dynamic nonhierarchical routing, which allows the network to route calls automatically over the best available path and manages the rapidly growing information loads, according to AT&T. Diverse routing will allow AT&T to skirt disaster that might otherwise shut a system down. The growing AT&T network will provide the capability for new services, allowing AT&T to plan and offer advanced communications services including wideband packet switching and offering compatibility with ISDN. Customers will have greater control over their systems generally, AT&T believes.

AT&T has had some success developing its "campus of the future," which used the Information Systems Network developed by what was AT&T Information Systems. The high-capacity, high-speed LAN provides voice, video, and data transmitted simultaneously.

In one such installation for the University of Pittsburgh, AT&T installed 350 miles of optical fiber on campus, which included 120 ports. Professors can store and tape lectures, to be called up at the appropriate time; students are able to access information about events from various kiosks located around the campus. Other services are to include accessing information from the university's 18 libraries via host computer hookup and campuswide electronic mail and video bulletin boards.

By late 1986, KMI estimated that dozens of colleges were moving to fiber to help with their requirements, including Rutgers, Ohio State, DePaul, the University of Alabama, Northwestern, Notre Dame, and Harvard.

To better prepare for the coming of fiber optics to the local loop, AT&T officials have urged companies to prewire new buildings, so-called "smart buildings," with optical fiber. (See Figure 3.)

The Regional Holding Companies have also recognized this coming market, although they are prevented by Judge Greene's decree from actually manufacturing LAN equipment. One example of this is the 50% ownership by BellSouth of FiberLAN, what was formerly a wholly owned Siecor subsidiary. FiberLAN's stated goal is to sell fiber optic based LANs. The fact that FiberLAN was already located within BellSouth's area and that BellSouth has been the most bullish of the RHCs when it comes to installing fiber gave the deal further credence.

Ameritech has purchased the rights to fiber optic based

Figure 3. New "Smart buildings" are being wired with fiber optics. Here optical fiber cable is supplied in 200 foot lengths and secured at 18-inch intervals with hold-down tape. (Photo courtesy of AMP Incorporated, Harrisburg, PA.)

LAN technology developed by Aetna Telecommunications Laboratories.

AT&T requested that the Justice Department make sure that neither BellSouth nor Ameritech manufactured LANs; each has been restricted to designing LANs for customers only.

In 1984, Southwestern Bell installed a LAN in its corporate headquarters in St. Louis, which it called "the largest LAN in the world under one roof." Much of the LAN was interconnected with optical fiber, so much so that Southwestern Bell called it "the most amount of fiber installed in one place." The Southwestern Bell LAN incorporates approximately 144 km of fiber cable carrying data communications to and from more than 2000 terminals and workstations in the 44-story building.

The U.S. Navy claimed that a fiber optic LAN being installed at the Port Hueneme Naval Ship Weapons Systems Engineering Station in late 1986 was the largest all-fiber optic LAN in the world. Designed by Applitek Corporation, the network initially connected between 500 and 600 devices in 30 buildings, with 2.5 miles of optical fiber cable comprising 243,000 fiber feet.

Other companies such as Fibercom, Fibronics International, TRW, and Astranet (which has a Xerox Corporation affiliation) have also come to market with fiber optic based LANs. In all, some 500 companies in the United States have the capability to introduce a fiber optic based LAN, calculates marketplace prognosticator Herbert Elion, of International Optical Communications. Of that group, Elion said perhaps 100 may actually offer a product.[8]

Meanwhile, customers were finding ways to match the advantages of fiber optics for their own data requirements. As mentioned, the Iowa based Norand Corporation has developed a data system connecting a cash register with a data storage unit via optical fiber, which in turn is connected to an off-site computer via traditional communications. Customers include 7-11

[8]"1986: U.S. Fiber Market to Reach $1 Billion Mark; Bypass to Become A Reality," *Fiber Optics News,* January 13, 1986.

Stores, Burger King, Long John Silvers, Pizza Hut, and Winchell's Donut Shop.

A secondary definition of a Harvard MAN refers to a metropolitan area network installed at Harvard University which is largely fiber optic based. The Harvard College observatory, Aiken Computation Laboratory, and Massachusetts General Hospital were the first three buildings to be tied together with fiber. The Harvard Medical School was also to be included.

As a matter of fact, college campuses and hospitals are two leading customers for fiber optic LANs, as the ability of each to operate is largely dependent on receiving information rapidly from a variety of different sources within its own complex. Others fibering their local environments include the Campbell Soup Company and Rockefeller Center.

While Kessler figured that the data communications market in terms of optical fiber was only one-tenth the size of the telecom market in 1986 (128,000 fiber km compared to 1,275,000 fiber km), the data communications market was growing substantially while the telecom market was not. The data communications market grew from $120 million in the United States in 1985 to $185 million in 1986 and $280 million in 1987, according to Kessler. That is expected to leapfrog to more than $1 billion by 1995.

"The reasons for this growth lie not only in fiber optics and its ability to economically carry increasing amounts of information, but in the economics of networked computer systems and the need to interface high-capacity data networks with telephone and video systems," according to Kessler.

The Commerce Department also found there is "enormous potential" for fiber optics in the computer field. "Fiber optics could become as important a building block for information technology as the semiconductor."

The all-optical broadband network of the future is a topic of increasing prevalence for communications engineers. At the International Symposium on Subscriber Loops and Services in 1986, there was general agreement that the broadband network would be an overlay network based on single-mode fiber. There was also the understanding that fiber component prices must come down considerably before that can happen. Still, the day is coming when a smorgasbord of services will be at the user's fingertips.[9]

"The office of the future is designed for an increased usage of data-processing equipment to provide stored reference data for the day-to-day operation of the business; to have analytical capabilities; to assess stored information at information banks, to do word processing; to distribute documents locally and remotely; and to communicate in real time with other offices, with telephone, data, and television in a multiply connected fashion," Kao observed in 1982.[10]

Optical technology continues to get closer to the mainframe computer itself, just as fiber optics continues to get closer to the home. Knowledgeable sources predict that the input/output of mainframe computers will be largely wired with optical fiber cable by the early 1990s.

IBM is not the only company seeking to improve the basic building blocks—the chips themselves. Advanced integrated circuits are expected to lead to higher speed computers, allowing LANs to reach speeds as high as 2.4 Gbps, and potentially much higher.

While optical computers will probably not need wiring as we now know it, very advanced lasers and photodetectors weaned on fiber optics will be the components. Thousands or even millions of these lasers and detectors could be used in array format, reaching terabit levels. Not surprisingly, Bell Laboratories has already established an Optical Computing Department, aiming for a full-scale model by 1990. Such optical computers could be the basic tools of the Strategic Defense Initiative, perhaps a major reason why the Soviet Union is also avidly interested in optical computing. Whatever the outcome, today's mainframes seem destined for obsolescence.

---

[9]"The All Optical Broadband Future," *Communications Systems Worldwide*, November 1986.

[10]*Optical Fiber Systems: Technology, Design and Applications*, by Charles Kao, New York: McGraw-Hill, 1982.

Chapter 11

Refitting the U.S. Military

THE UNITED STATES military fully understands the importance of optical communications. Optics, particularly fiber optics, has reached buzzword status in the U.S. Armed Forces, meaning that a military application or project incorporating the new technology might be approved, while one using more conventional technology might not. From early examples such as the use by U.S. Navy signalmen of short and long flashes of light to communicate between ships, to more recent plans to rewire entire ships and airplanes with fiber, to the planned deployment of large amounts of fiber in unmanned battlefields of the future, the military is gradually moving its communications and information requirements from electronics to optics. The military has been an interested observer of and participant in the growth and development of fiber optics from the time the technology was born. The advantages of fiber optics are "key to improving the strategic and tactical capabilities of the military forces," according to Charles Kao.[1]

Fiber optics played a limited, yet very effective, role in the

---

[1] *Optical Fiber Systems: Technology, Design and Applications,* by Charles Kao, New York: McGraw-Hill, 1982.

Vietnam war. It was used in passive surveillance devices developed by the Army Night Vision Laboratory both from the ground and the air. In one instance, the night vision equipment was responsible for locating approximately 200 Vietcong soldiers just prior to their planned nocturnal attack on a town. The raid was successfully thwarted as the result.[2]

By the time of the first major fiber optic conference, OFC '75, it was clear that all of the major armed services had found out about fiber optics and were experimenting with it. This was demonstrated by a paper presented at the conference by representatives of the Army (Army Electronics Command's L. Dworkin), Navy (Naval Electronics Laboratory Center's D. Albares), and Air Force (Air Force Avionics Laboratory's K. C. Trumble) entitled "Prospective Applications for Fiber Transmission in the Military."[3]

The paper was important not only because it demonstrated an intelligent interest by the military in fiber, but also because it pointed to a number of applications that would come into effect a decade or two later. The authors began the paper by saying that "emerging fiber optics technologies promise to have a large impact on Department of Defense communications and information processing in both near- and far-term applications." An early market they identified was the sending of information within an airplane, where fiber optics offered "an ideal solution" for "providing the necessary immunity and eliminating the risk of malfunction or serious damage due to lightning, static discharges, electromagnetic pulse (EMP), etc."

Using fiber optics to wire ships and submarines was also presented as a desirable and achievable task. "Internal shipboard

[2] "Passive Vision Aids Pierce Vietnam Night," by B.M. Elson, *Aviation Week*, June 3, 1968, pp. 89–90.

[3] "Prospective Applications for Fiber Transmission in the Military," by D. J. Albares, Naval Electronics Laboratory Center, San Diego; L. V. Dworkin, Army Electronics Command, Ft. Monmouth; and K. C. Trumble, Air Force Avionics Laboratory, Wright-Patterson Air Force Base, Ohio. Paper presented at OFC '75.

communications suffer from many of the aircraft problems. . . .",
the paper said. Fiber optics could help to eliminate the "serious
problem" of electromagnetic interference (EMI) within a ship. By
the time the paper was presented, fiber optics was already being
used for communications on board the USS Little Rock.

Undersea fiber optic systems also presented "substantial
advantages" over electromechanical cables, which suffered size,
weight, attenuation, bandwidth, radio frequency interference
(RFI), and cost limitations." Potential fiber optic applications
included acoustic arrays, tethers, and telecommunications.

Land based uses included strategic and tactical communications, in addition to long-distance telecommunications. Important features of fiber optics for these applications included freedom from EMP, reduction in weight and size, freedom from
signal leakage, low signal loss allowing for less repeaters, and
immunity to RFI and EMI.

Clearly, the military was going to satisfy its own specialized
requirements by matching them with the unique characteristics
of fiber optics. By the mid-1980s, hundreds of millions of dollars
were being spent annually. The Pentagon had even begun rewiring itself with fiber optics.[4]

There are estimated to be 149 applications of fiber optics in
the military in the 1986–1990 time period, according to Richard
Schade, whose four-member group at the Defense Electronics
Supply Center (DESC) is responsible for developing standards
for using fiber in defense applications. Some advantages of
fiber optics to the military that DESC identified included
greater bandwidth, smaller size, lighter weight, lower attenuation, no EMI, signal confinement, safety, and lower cost.
There are also no short circuits, arcing, or sparks when bare
conductors touch. A number of fiber optic applications in
the military have resolved difficult problems in design and
test, according to DESC. "In some cases, fiber optics have

---

[4]"Pentagon to Be Rewired With Fiber, If Xerox Can Work Bugs Out of GSA
Contract," *Military Fiber Optics News*, May 23, 1986.

tied together heavily loaded data networks in environments where even one error caused by nearby RFI can foul up a test or data system." In recent years, "an increasing number of test instrument designs have used fiber optics to handle digital or analog for some of industry's most sophisticated instruments."

Defense Department spending on fiber optics reached $186 million in 1986 and was expected to grow to $268 million by 1987, according to independent consultant George Rappaport. It is expected to grow to $333 million in 1988, $412 million by 1989, $509 million in 1990, and $625 million in 1991.

"It is my opinion that eventually all firms and government agencies will use fiber optics," Rappaport told the Kessler conference in October 1986.[5]

There are certain broad programs in which the various military agencies have to work together. One is in ensuring that a communications system remains operating in case nuclear weapons are used. If that unwelcome event were to occur, optical fiber, which is dielectric (nonelectric), would have major advantages. Researchers, however, must first develop a way of developing radiation-resistant fiber (rad-hard fiber). The central problem has been to create a fiber that—when exposed to radiation—will not darken; this increases attenuation and can knock a fiber optic system out of commission.

As the early OFC paper indicates, each of the services could use fiber. The advantages "are pretty much across the board," according to AFCEA-International President Admiral Jon Boyes, "whether it's in the cockpit, or the submarine, or the man that's out there on the ground."[6]

The Army is recognizing that fiber optics can be used to effectively link its military bases and has put a fiber optic network

---

[5]"Rising Markets for Military Fiber Optic Systems," by George Rappaport, presented at the 1986 KMI conference.

[6]"Admiral Boyes Cautions Would-Be Fiber Optic Military Suppliers to Exhibit Patience," *Military Fiber Optics News*, June 13, 1986.

throughout South Korea to accomplish just that. [This is the Korean Improvement Program (KIP).] The Army has also learned that it can tether a missile to a command center with optical fiber to fly it intelligently and land it precisely on a target. Fiber's lightweight advantages can provide good point-to-point interconnect on a battlefield. The fact that it weighs from 6% to 25% of what metallic cable does means that an infantryman can carry the cable in his backpack during an invasion.

The Navy is finding that wiring a ship with fiber optics can save the service weight and money. The Navy in 1986 began an aggressive effort to replace copper with optical fiber on its ships, with the belief that a decade or two later the ships would be mostly wired with fiber. For example, replacing copper with optical fiber can trim 160 tons off a two-level ship, according to James Davis, of Naval Sea Systems.

The Air Force should "substitute photonic devices for electronic devices wherever feasible to defeat electromagnetic pulse, radiation, and electronic warfare threats," according to findings released by Project Forecast II, an Air Force Think Tank.[7]

"The goal," according to Project Forecast II, "is to produce systems—like strategic or tactical battle management workstations—that employ photons instead of electrons to sense, compute, process, and transmit signals." The study predicted that "mastering photonics will require the integration of optical fibers, optical materials, optical sensors, and optical kill mechanisms, plus a significant investment in optical processing."

Other governmental agencies are also aggressively using fiber optics. NASA has been testing fiber since the mid-1970s and claims that the first field application of fusion splicing was performed in manholes at Kennedy Space Center in 1976.[8] (See Figure 1.)

---

[7]"Fiber Optics 'Revolutionary' Role in Future Air Force Seen," *Military Fiber Optics News*, June 27, 1986.

[8]"Pioneer Fiber Optic System Survives Harsh Environmental Conditions," *Corning Guidelines*, Vol. 2, No.3.

Figure 1. Top-down look of installation at Kennedy Space Center in Florida. The first fusion splicing was performed in manholes here in 1976, according to Corning. (Photo courtesy Corning.)

NASA already is studying the use of an FDDI-compatible LAN in the Space Station and adorning the Space Station with fiber optic sensor arrays to detect outside conditions.

Fiber optic data communications systems are also used by the Central Intelligence Agency, the National Security Agency, the Federal Bureau of Investigation, national energy laboratories (including Lawrence Livermore, Los Alamos, and Oak Ridge) U.S. embassies, the National Institutes of Health, and the General Services Administration.

Fiber optics can be used in ways that are common to both the commercial market and the military. Stringing fiber between military installations and secure campuses for computers are similar to point-to-point and LAN applications on the commercial side. However, there are ways the military is incorporating fiber optics into its own scheme of operations that differ from typical commercial applications. The use of fiber to tether missiles so a missile can be controlled from a command center falls into that category. The missile is guided to the target, where it is detonated. (see Figure 2.)

This is exemplified by the Fiber Optic Guided Missile (FOG-M) program, which has received enough military attention to command an $80 million budget in fiscal 1987 and reach the status as lead candidate for the Non Line of Sight (NLOS—which involves destroying targets beyond the view of the battlefield unit) program element for the Forward Area Air Defense program. (FAAD may receive $2.7 billion over the period of a decade.) Optelecom President William Culver, who left IBM Corporation to form Optelecom because of IBM's refusal to commit to the FOG-M technology, believes fiber guided missiles will "revolutionize" tactical warfare.

FOG-M brings a new accuracy to short-range missiles. Using FOG-M, optical fiber cable is paid out as a missile is launched and provides the vital lifeline between the missile and a control base on the battlefield. The system uses a bidirectional optical link connecting a nose-cone television camera in the missile with the individual controlling the missile. FOG-M provides a unique opportunity for the controller to see where the missile is going via the uplink and then guide it to a precise target area via a secure downlink. The operator may be housed in a tanklike armored missile carrier or on a mobile platform. The early missiles weigh approximately 70 lbs and fly at approximately 180 mph.

FOG-M has been under study since the early 1970s. The link, which has involved extensive research in spooling technology, has operated efficiently in its first 11 tries. The missile performed well in 7 of those 11 tests. "From our point of view, we've proven the concept of fiber optic guidance," observed Bob Coon,

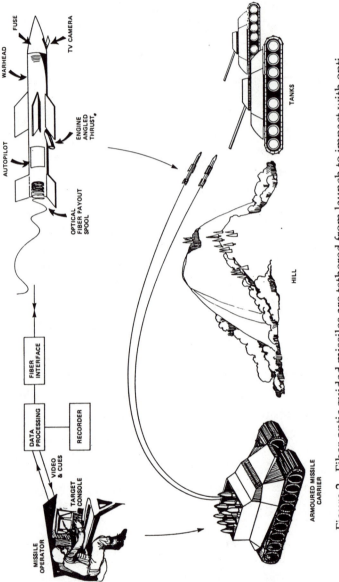

Figure 2. Fiber optic guided missiles are tethered from launch to impact with optical fiber cable. The missiles are controlled by an operator. As the diagram indicates, fiber is spooled out of the rear of the missile. (Photo courtesy Optelecom.)

a senior program manager at ITT Electro Optical Products Division. ITT and Hughes Aircraft have both played pioneering roles in the technology.

The precise missiles have the advantage of hitting a tank from the top, thus circumventing the heavy armor that pads the front. Culver also believes the missiles can be used against helicopters. The missiles are also light enough so that they could be carried by helicopters such as the AH-64 Apache, according to Culver.

Hughes is also fashioning lighter weight fiber guided missiles that go shorter distances than what is being proposed for FAAD. A miniature FOG-M whose goal is to travel from 2 to 5 km, this smaller missile is being proposed for the Advanced Antitank Weapon System-Medium (AAWS-M) program. The first flight for this lighter unit is slated for the first quarter of 1988.

Hughes is also proposing an intermediate design for a separate but related Army program, Advanced Antitank Weapon System-Heavy (AAWS-H), which is planned to replace the TOW-2 missile. This would have a range of 5 to 10 km, greater than the missile proposed for the AAWS-M program, but less than FOG-M, which has a range of 10 to 15 km. The cost of each of the three missiles is expected to be in the same ballpark, although the smaller units could be slightly cheaper, according to Jim Oddino, AAWS-M Program Manager for Hughes' San Fernando Valley division. (See Figure 3.)

While the plan has been to use fiber guided missiles for short-range applications, these capabilities could also be extended significantly, according to Culver. A spool housing up to 40 km has already been developed. The missiles could potentially be built for as little as $20,000 each, using off-the-shelf electronics and a ground-based computer system.

Optelecom has also received military contracts for designs where robots are linked by optical fiber and operated remotely. The robots can work in areas where conventional copper cable might run into problems, such as nuclear power plants. Other potential applications include terrorist situations, unexploded

AAWS-M ENGAGEMENT SCENARIOS

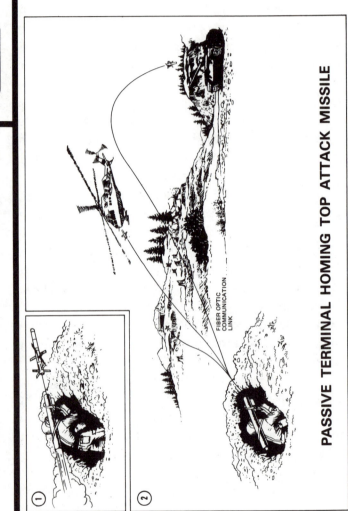

PASSIVE TERMINAL HOMING TOP ATTACK MISSILE

FIBER OPTIC
COMMUNICATION
LINK

Figure 3. Guided fiber optic missiles are also being proposed for deployment in systems lighter than FOG-M. AAWS-M deployment could use missiles operated by an infantryman on the ground from distances of 2 to 5 km for various applications. (Photo courtesy Hughes Aircraft Co.)

weapons retrieval, firefighting, combat work, and toxic waste handling, according to Culver.

Remotely operated vehicles are playing a key role in improving surveillance for both tactical and scientific purposes. The U.S. Marine Corps is developing the Advanced Remotely Operated Device, dubbed "the flying dixie cup," a gasoline-powered unmanned surveillance vehicle for tactical situations. The Navy is exploring the use of unmanned submersibles launched from submarines and tethered with fiber that can comb the ocean's floor, literally allowing human beings to view things heretofore unseen.

The military has also found that fiber optics has the potential to simplify mechanical tasks. The Advanced Digital Optical Control System (ADOCS) uses fiber optics to better guide helicopters. The Army has dubbed the process "Fly By Light." In this process, optical fiber is used to send commands from the pilot's control stick to the primary flight control system (PFCS) in the helicopter via an electro-optical interface. The PFCS consists of the mechanical devices that control the helicopter's actions. A second flight control system is used as a standard backup. Parallel optical fiber signal paths connect to the backup system's processor, known as the automatic flight control system. Both control systems are, in turn, connected via optical fiber to the power actuators, which physically move the aircraft's control mechanisms. Optical fiber is also used as a data bus connecting to the helicopter's various cockpit information displays.

The ADOCS control transmission system is far simpler to use than a dual-mechanical system. It replaces 150 bell cranks and push rods with six optical fiber cables and eliminates the mechanical system's rigging points.

ADOCS is being developed with the UH-60 Black Hawk, AH-64 Apache, CH-47 and Light Helicopter Experimental helicopters in mind, according to Jerry Ervine, a spokesman for the Army's Aviation Systems Command. The Army awarded Boeing Vertol Company a $25 million contract in 1981 to prove that a helicopter can be controlled by commands sent along optical fiber, using a Sikorsky UH-60 Black Hawk helicopter as the test vehicle.

ADOCS has the potential to be selected as the flight control system for the Army's proposed $40 billion LHX program, a contract that could lead to its inclusion in other aircraft. The Army, in fact, is hoping that the development of ADOCS will allow aircraft control systems to enter a new generation, "making mechanical control systems obsolete," replacing them with a system that is less vulnerable to mechanical stresses and other intervening factors. It is the inherent characteristics of fiber optics that the Army believes will provide a significant stride beyond these other systems, reducing a flight control system's vulnerability to RFI, EMI, and nuclear effects.

The Army is also interested in using an optical fiber link in its Army Advanced Attack Helicopter as a medium to carry a light source to an internal viewing device in the cockpit and to supply light for an alphanumeric LED in a head-down display used for target acquisition. The process is called Target Acquisition Designation Site/Pilot Night Vision Sensor.

The various advantages of fiber optics to provide for longer repeaterless distances, immunity from RFI and EMI, operation without emitting a radio frequency signature, and lightweight and smaller space make it a natural for battlefield operation. This is demonstrated in the Fiber Optic Transmission System-Long Haul (FOTS-LH) program.

The Army Communications-Electronics Command initiated FOTS-LH in 1982 to develop a system that eventually could replace all of the military's CX-11230 twin metallic tactical coaxial cable with optical fiber cable. The hope was to create a fiber optic transmission system that could become the new workhorse tactical communications system throughout all of the services, according to Paul Smith, FOTS-LH project engineer at ITT Defense Communications Division, which is the prime contractor.

The Army originally proposed transmitting signals to 64 km, but ran into problems, including how to design and include the repeaters necessary in such systems. The program has been tailored to repeaterless links running to 8 km using transmission rates of 4 Mbps and 18 Mbps. Repeaters may eventually again be tried to provide for longer distances.

The Army Communications Electronics Command sent out the first phase of the FOTS-LH solicitation, the requests for technical proposal, on March 16, 1987. There were actually two separate solicitations, one for the cable (Fiber Optic Cable Assemblies, including bulkhead connectors and permanent cable repair sets) and one for the electronics (Remote Optical Assemblies, including field test sets).

If funding for all three years of the program is approved, as many as 15,000 cable assemblies could be required (costing up to $5,000 per assembly) and as many as 5600 transceiver units could be requested. (These are to consist of edge-emitting LEDs operating at 1300 nm wavelengths and pin-fet receivers).

One of the largest military programs is ARIADNE, which is being studied by the Navy. ARIADNE could eventually protect the coastal waters of the United States by sensing the presence of enemy submarines.

The Pentagon began testing a prototype version of ARIADNE in 1986. The undersea network is made up of hydrophones, which are basically undersea microphones, strung together with optical fiber cable and connected to a control center, which can be located either on shore or aboard ship. Pentagon officials reason that optical fiber cables are lighter, cheaper, easier to install, and have much greater transmission capacity than their counterparts. The greater transmission capabilities and immunity from interference allow the quality of the data to be much better than it was using conventional systems. Quality is a key concern when one considers the sensitive sonar signals required to first sense and then identify a submarine.

"This program has extended the technology of fiber optics to strategic or tactical or undersea surveillance," according to Dr. Robert Duncan, who headed DARPA when ARIADNE was an experimental program in that shop. During 1986, repeater/multiplexer/connector prototypes were in the final stages of production.

Both SpecTran and Corning hope to provide optical fiber for ARIADNE. High-grade, high-strength, small-diameter, single-mode fiber is required. Lockheed Advanced Marine Systems has developed a low-loss, wet-mateable single-mode connector for ARIADNE. TRW's Electro-Optics Research Center has provided single-mode optical fiber rotary joints for ARIADNE.

Because enemy vehicles would probably attempt to tear up the cable when they locate it, the Navy will probably have to replace it on an annual basis, meaning that large amounts of the optical fiber may be required.

The Naval Air Systems Command Center is incorporating an optical fiber data link into the AV-8B Harrier, a single-engine, single-crew-member attack aircraft with vertical/short take-off and landing capabilities. The optical fiber link would interface with an optical disk storage unit with a digital display map generator. First flights using the optical data link are slated for May 1987, with production by 1989.

The Navy is also considering the use of fiber for an operational prototype airship, which could eventually be deployed in a 48-airship fleet for early warning for battleships. Single-mode fibers on the 420-foot dirigible would connect all sensors, flight control computers, and actuator drive units, according to E. J. Carpenter, Airship Industries' project manager for optically signaled flight control systems. Fiber was chosen to reduce the threat of lightning strikes and to ensure signals without interference, according to Carpenter.

The Air Force is studying the use of optical fiber cable for the purpose of launching ground based cruise missiles and intercontinental ballistic missiles by using a missile laser ordnance initiation (LOI) system. Hercules' LOI system received an $84 million contract from the Air Force Ballistic Missile Office to develop the firing system, which will replace more conventional firing systems.

The Air Force is also developing ways to use optical fiber to extend the distance between a remote radar site and operating center in a number of different programs, both commercial and tactical. The need for this was graphically illustrated, according to Pentagon officials, when a U.S. remote radar site was fired on by Libyan forces in 1986.

Under the AN/TPS-43E Fiber Optic Radar Remoting Kit (FORRK) program, radar kits being developed have used ITT Corporation's T-2500 optical fiber and trilevel receivers and transmitters from Honeywell Inc. to provide a 12,000 foot link to remote radar. This extends previous cabling distance significantly.

These and other military programs can add up to millions of dollars for savvy companies providing the necessary products. A number of U.S. businesses are attempting to develop an expertise in the military fiber optics market.

Corning has a longstanding relationship in providing its optical fiber to the military and, in 1986, created an Advanced Fibers Products Division, whose mandate is to focus on specialty fibers and fibers for use in the military/government market. Siecor decided the same year to build a specialty cable manufacturing plant in Hudson, North Carolina, near its facility in Hickory to (among other things) manufacture ruggedized, aviation, and undersea cable for the military market.

Continuing its long-term relationship with the military, AT&T has been selected to provide the Defense Commercialization Telecommunications Network, a 10-year project costing more than $400 million. The job is to interconnect 15 major military nodes throughout the United States. Fiber optics will be involved with connecting the 150 to 160 military bases spread throughout the country with those nodes. Satellite is playing a major role in the network.

The federal government is also involved in a big bucks program to provide for its civilian communications needs via the FTS-2000 program. The major fiber carriers are joining with government contractors, including Martin-Marietta, to try to win the contract. Much of the traffic will undoubtedly end up over fiber, should the system be successfully contracted.

AT&T's approach generally to providing fiber optics to the military is three-tiered, according to Charles McQueary, director of undersea systems development at Bell Laboratories. The formula includes capitalizing on experience gained on the commercial side, instead of reinventing the technology; building military systems from conventional or militarized conventional components, because the development expenses have already been assumed by the company; and taking commercial designs and adopting them for military use.[9]

AT&T is manufacturing specialized optical fiber cable for the military and has developed a militarized data link based on its commercial ODL 200 fiber optic data link.

SpecTran Corporation has also tried to focus its efforts to take advantage of the military fiber optic marketplace, hoping to provide specialty cable. This effort has intensified since SpecTran came up with major problems in trying to get into the commercial market.

Frost & Sullivan estimates that 80% of the military fiber optic market in the late 1980s will be in providing optical fiber cable.

One problem that U.S. fiber optic suppliers on both the commercial and military sides face is attempting to offer state-of-the-art fiber optic products abroad. This is the result of the military's effort to keep U.S. fiber optics technology out of the hands of enemy countries. Armored optical fiber cable is categorized by the U.S. government, for example, under "arms, ammunition and implements of war" and its export comes under strict control.

Admiral Boyes has characterized the Soviet Union as possessing "damn good fiber optics." The recent bugging of the new U.S. embassy in Moscow provides vivid proof of that.

---

[9]"AT&T's McQueary Outlines Company's Strategy, Development Aimed At Military Fiber Market," *Military Fiber Optics News*, September 5, 1986.

While there is money to be made in the military fiber optic marketplace, many suppliers find that gaining a foothold is not always easy, despite Pentagon assurances that the military is looking for "off-the-shelf" products. This is particularly true for suppliers who are used to dealing with the less cryptic ways of the commercial marketplace.

"When you talk to many fiber optics people, they do not understand the military and government," Boyes observed. "They don't want to get caught up in a contracting process, they like quick turnover." But, Boyes added: "The more farsighted companies are saying 'There is a big, big market for us if you know how to get into it. . . .'"

The future of fiber optics in the military is very bright. Photonics has the potential of revolutionizing "the way military tacticians think about strategic, tactical and space warfare," according to Project Forecast II.

One potential Air Force application is to build "smart skins" on an aircraft's wings, housing embedded phased arrays allowing aircraft to sense and communicate in optical and other frequency bands. "Smart skins" are considered "remarkably survivable" in flight and could also enhance an aircraft's ability not to be detected in flight.

The Navy has taken the lead role in developing what will be the next generation of optical fiber, fluoride based fiber. While the U.S. terrestrial market appears to be well served by silica based fiber, undersea applications present a brave new world, and this is where fluoride fiber could come in handy. Fluoride fiber has the potential for reaching losses of 0.003 dB/km, which could dramatically increase the distances that fiber signals can travel without boosting.

The Navy in 1986 announced a 30-month R&D program on fluoride fiber, whose goal is to demonstrate "the feasibility of long-length fluoride optical fiber data links." While early applications might go for such projects as ARIADNE, the potential for telecommunications is certainly there.

Fiber optics will play an "absolutely essential" role in the Strategic Defense Initiative, according to Boyes, if SDI goes forward as proposed by the Reagan administration. Included will be the various supercomputers required for SDI. SDI Optical Computing Chief John Caulfield is already talking about sending information in 10^{14} streams—levels three orders of magnitude higher than current practices. This will be possible only through optical communications.

One thing is certain. The battlefields of tomorrow will incorporate fiber optics for a wide variety of purposes. The military will never again go about its business as it has before.

Chapter 12

"The Oceans will be Littered with Fiber..."

D R. CHARLES KAO had just given a paper at the plenary session of OFC '83 in New Orleans suggesting that some day repeaterless spans of fiber optic cable stretching the lengths of oceans would be possible.

In the dark of the reception area outside the main convention hall, he was reminded that it would be at least five more years before even the first transoceanic fiber optic cable was slated to come into being. Given those circumstances, the Chinese-born fiber optics pioneer was asked if such a statement was not a little premature.

Kao at first acknowledged that "it would be well into the twenty-first century" before such a repeaterless oceanic system would come about. But then he looked his inquisitor directly in the eye and quietly noted: "By that time, the oceans will be littered with fiber."

There are three spectacular achievements that have driven submarine cable technology to where it is today, according to a study by the U.S. Commerce Department's National Telecommunications and Information Administration (NTIA).[1]

[1] 1984 World's Submarine Telephone Cable systems (GPO 003-000-00636-1), National Telecommunications and Information Administration.

The first was the installation of the initial successful transoceanic telegraph cable from Ireland to Newfoundland by Cyrus Field and Dr. William Thomson (who later became Lord Kelvin) using the Steamship Great Eastern in 1866. The second was the installation of the first transoceanic telephone cable, TAT-1, in 1956. This was a cooperative enterprise involving the United States, Canada, and Britain. The third event, which is ongoing, is characterized by NTIA as the "fiber optics revolution." Due to fiber optics, the transoceanic communications scene has "changed dramatically," according to NTIA.

The potential use of fiber optics in undersea applications became apparent in the 1970s. Remember that the early applications costed in for long-haul communications, that fiber could more economically send a signal longer distances because it required less repeaters and it provided greater transmission capacity. This was music to the ears of those developing undersea communications systems, who were dependent on placing repeaters every mile or so on conventional copper cables.

Since the advantages were known early on, one might wonder why it will take until the late 1980s before the first transoceanic fiber optic network, TAT-8, will come about. There are key differences in building a system on land and building one that rests on the ocean's floor.

"Since a deep-water failure results in high repair cost and a huge loss of revenue, an undersea system must be made orders of magnitude more reliable than a terrestrial system," stated a paper presented by Bell Laboratories researchers Peter Runge and Patrick Trischitta. The goal is not to have more than three cable repairs over a 25-year period.

That is an impressive objective when one realizes what could happen to a cable. One of the main culprits, for example, has been trawler damage. If optical fiber does have a mortal enemy, it is water, so protection takes on a special significance. The cable must also be very strong to withstand ocean water pressure at depths to 18,000 feet and to maintain its integrity for the times

when it will be hauled from the ocean's floor to be t
Figure 1).

Fortunately, the designers of undersea fiber optic
have learned much from earlier systems. For exampl... ...arlier
designers had to deal with a variety of potential obstacles,
including making sure that cables do not collapse under pressure, are watertight, can withstand tension resulting from laying
and recovery activity, and are corrosion resistant. The electrical
part of the cable must also penetrate the bulkhead of the hous-

Figure 1. Prior to actually placing optical fiber cable on the ocean's
bottom, AT&T performs significant ocean tests. In this photo, Al Quigley from AT&T Bell Labs examines optical fibers emerging from a pressure vessel in which engineers simulate undersea conditions. The
photo was taken in 1985, three years before the first transoceanic system was slated for cutover. Quigley is a member of the Undersea Cable
and Apparatus Design Department. (Photo courtesy AT&T Bell
Laboratories.)

ings. Fiber optic designers are borrowing many of these features when building fiber optic undersea links.

One specific problem that fiber optics has introduced is splicing. Fiber splicing must be far more accurate than the conventional coaxial procedure of joining robust copper conductors together.

One rule of thumb is that it will take 10 years before a new terrestrial based cable technology can reliably be used on the ocean's floor. That seems to be about right in the case of fiber optics, as well.

It certainly was no surprise that AT&T would be spurring efforts in the United States for such undersea systems, given its experience with fiber optics and its early work with undersea transoceanic communications systems. Perhaps more surprising was the level of activity from companies in other countries interested in building undersea fiber optic networks. Various countries were getting experimental undersea fiber optic systems operational as soon as possible. (See Table 1).

Despite the fact that fiber optics is "revolutionary" in its undersea applications, the entities installing it around the world are not. Rather, the same countries who suffered through the submarine testing of cable and electronics years ago are (for the most part) now positioned to take advantage commercially of the new benefits fiber optics can bring.

In the United States, throughout the telegraph cable era, only Simplex Wire and Cable Company produced undersea cable. AT&T took control of the remainder of the system, and the two have continued to work together. The U.S. Navy has also aggressively undertaken an undersea program, using fiber for telemetry, sensing, and communications.

The many manufacturers of the telegraph era in Great Britain were consolidated finally into Submarine Cables Limited, which was in turn taken over by Standard Telephones and Cables in 1970. STC, which had an early interest in fiber optics, was the sole producer of submarine cable in the United Kingdom by the mid-1980s. STC and its antecedents have aggressively installed

Table 1. Fiber Optic Experimental Systems. STC, in cooperation with British Telecom International, was credited with operating the first experimental undersea fiber optic system, followed in hot pursuit by the Japanese. (Source: NTIA).

Country	U.K.	Japan	Japan	Japan	Japan	U.S.A.	France	France	Spain	France
Entity(ies)	STC BTI	NTT	KDD	KDD	NTT	AT&T	CNET CGE	CNET CGE	CTNE AT&T	CNET CGE
Date	1980	1980	1981	1982	1982	1982	1982	1984	1985	1985
Location	Loch Fyne	Izu Peninsula	Sagami Bay	Sagami Bay	Sagami Bay	500nm ENE of Bermuda	Mediter-ranean	Mediter-ranean	Canary Islands	Mediter-ranean
Landing Points	Loop	Inatori Kawazu		Ninomiya Loop	Yahatano	Ship Loop	Cagnes-Sur Mer, Juan-Les Pins	Port Grimaud, Antibes	Las Canteras Las Calletilas	Marseille Ajaccio
Length of cable, km	10	10.2	4.5	50	45	18.2	20	80	104	400
Cable Design	4 MM 2 SM	5 SM 5GI	6 SM	6 SM	4 SM	12 SM	2 SM 4 MM		6 MM	2 fiber pairs
Water depth meters	Shallow	200	500–1500	1300	1000	5000	1000	1300		2500
Number of Regenerators	1	none	1	2	2	2	none	2	4 + 2	8
Filter Mode	SM/MM	as above	SM	SM	SM	SM	as above	SM		SM
Number of pairs	3	as above			2	6	3	2		2
Wavelength microns	1.3	1.3	1.3	1.3	1.3		1.3	1.3		1.3
Transmission Rate Mb/s	140	6.4, 32 100	280	300	400	274, 420	34	280	295.6	280

fiber optic undersea cable, with British Telecom International playing a support role.

In France, Cables de Lyon has enjoyed total eminence in undersea cabling from the beginning and landed all cable on French soil until 1979. Cables de Lyon is now a major partner in the French Submarcom consortium that has control of France's involvement.

The Japanese made their entry relatively late in the business and, as is not totally uncharacteristic, significantly used other previously developed technology in their initial design. Nippon Telegraph and Telephone (NTT) has handled all domestic installation, while Kokusai Denshin Denwa (KDD) is the international arm. This is changing, however, as the Japanese also try deregulation. NTT and KDD competitors are beginning to crop up.

KMI's Robert Holtzman calculated that there were 11 companies marketing submarine cable by the close of 1986. Six are marketing repeaterless and/or branching units. The engineers designing and building the undersea systems have learned their lessons well. Generally speaking, they take technology that has already been working effectively on terrestrial systems and adopt it for undersea applications. The early repeater designs, for example, were built with a design objective of 20 years. Most have achieved the goal.

One of the lesser-threatening problems has been that of route location. The general rule is that there are very few places where you cannot physically lay a cable, although the NTIA report points out that certain locations are more desirable than others. Some considerations include geography of the coastline, prevailing currents, depth contour, condition of the sea bed, nature of surface and underwater traffic, and distance to metropolitan areas.

The first sea trial of a commercial fiber optic telephone cable system took place at Loch Fyne, Scotland, in February 1980. The venture involved STC and BTI.

The Loch Fyne experiment was followed by five prototype systems by the Japanese in 1980–1982, three by NTT and two by KDD. AT&T's first experimental system was tested in 1982, as was the first by the French consortium Submarcom. STC tested a system in 1983, and Submarcom tested a system in the Mediterranean in 1984.

The United States military, which had been exploring the possibilities of undersea fiber optic systems from the 1970s, completed installation of a 150-km repeaterless undersea system in 1985 with the help of AT&T. The low-capacity system—operating at 3.088 Mbps—was placed in service as a data link for the remote Air Combat Maneuvering Instrumentation range. The ranges train air crews in Air Combat and Tactics Development using actual jet fighter aircraft and electronically simulated weaponry. The system is notable in the distance that it sends repeaterless signals.

A big booster of fiber optics has been George Wilkins, formerly of the Naval Ocean Systems Center in Hawaii. "We are very close to going out in a modified fishing boat and paying out 20,000 feet of fiber optics cable from a reel," Wilkins noted in 1985. "And that is entirely due to the advantages of fiber optics." Wilkins told a Marine Technology Society forum that you could connect San Francisco and Tokyo with fiber by loading a tuna boat to two-thirds of its capacity.[2]

Wilkins has become associated with the Dumand Array, an international scientific project to lay fiber cable strings straight up about 1 km from the ocean floor. The scope of the project is to find out more about neutrinos, uncharged elementary particles said to be massless. Neutrinos give off light as energy. The project has also received funding from the Soviet Union, France, the National Science Foundation, the Defense Department, and Mobil Corporation.

Optical fiber has also been used as part of a composite tether cable for remotely operated unmanned submersibles—vehicles used to inspect and repair underwater structures and explore

---

[2]"Fiber Optics: Sitting in The Catbird Seat of Marine Communications Technology," *Fiber Optics News*, March 4, 1985.

ocean environments. Normally a cable will transmit electrical power from the ship above to power the vehicle, and optical fiber will be used for video signals and data transfer. A major application is for oil exploration.

The STC-BT consortium was credited with constructing the world's first international undersea optical fiber cable in 1985. The route traverses 122 km and connects Broadstairs in the United Kingdom with Ostend, Belgium.

The cable houses three pairs of optical fibers, each operating at 280 Mbps, providing for 12,000 simultaneous phone calls, in effect doubling the capacity between the United Kingdom and Northern Europe. The route employs three fiber optic repeaters and is being funded by Belgium, the Netherlands, West Germany, and the United Kingdom, which will own 50%. But STC has admitted that the $10.15 million it received will not cover expenses and that it is operating at a loss. However, the experience gained from the system made it worth the effort, STC reasoned.

In autumn 1985, AT&T completed a 120-km undersea fiber optic network joining the Canary Islands of Gran Canaria and Tenerife off the Spanish coast. AT&T considered the network its prototype TAT-8 link, even though the system was carrying commercial traffic for the Spanish National Telephone Company. AT&T considered the system the first deep-sea commercial fiber network, reaching depths of 3000 meters.

It took the C.S. Long Lines—AT&T's premier cable laying vessel—three days to install the cable. Long Lines, which is also expected to take the lead in the TAT-8 installation, has been installing cable since 1963. The ship has an overall length of 511 feet and a beam of 69 feet. After testing the system with two repeaters, Bell Laboratories engineers intentionally broke the cable and spliced in a third repeater, to help demonstrate that TAT-8 will be recoverable for repair, if necessary. (See Figure 2.)

The system has already located a problem that was not recognized as a potential area of concern. At three different times during the first nine months, the cable was attacked by sharks,

Figure 2. The C.S. (Cable Ship) Long Lines is now used for laying optical fiber cable, but has in the past laid more undersea copper cable than any other ship in history, according to AT&T. It is expected to soon install 3145 nautical miles of fiber optic cable across the Atlantic and more than 10,000 nautical miles of fiber cable across the Pacific. (Photo courtesy AT&T.)

and shark teeth were found embedded in the cable. The attacks were made approximately three-fourths of a mile below the surface, where visibility is almost nil. The sharks were estimated to be from 1 to 3 feet in length.

AT&T Communications manager Carl Jeffcoat theorizes that the sharks must have been attracted to the electric field generated by the cable. The reason the sensory-laden sharks were attracted to optical fiber cable—and have not been enticed by the more conventional coaxial cable—was that no outer shielding was applied to surround the electrical field, a condition that AT&T can correct. Bell Labs can simply shield the outer part of the cable, where necessary.

The prototype was helping to clear the way for TAT-8, the first transoceanic fiber optic system in the world. TAT-8 has come to symbolize many things. To some of the pioneers, it represents the coming of age of fiber optics, tangible proof that fiber optics is for real and that it is affecting the world community.

The 28-member consortium controlling the development of TAT-8 awarded AT&T the major share of the contract, with Simplex Wire and Cable providing the cable. The consortium decided to split the job among the three major contractors, with AT&T installing 3161 nautical miles and receiving $250 million, Standard Telephones and Cables receiving $52 million to install 280 nautical miles, and Submarcom awarded $33 million to install 166 nautical miles. (See Figure 3.) A major factor in the decision was that AT&T could insure its portion of the job for 10 years, far longer than either of the other two organizations. AT&T also reportedly offered the lowest bid.[3]

The route configuration features an offshore branching design, which brings TAT-8 from Tuckerton, New Jersey, to Widemouth, England, and Penmarch, France. The single-mode fiber system uses two active fiber pairs with a capacity of 280 Mbps. A separate spare fiber pair is included, if needed. The system uses 125 repeaters, an average of one every 31 miles. Integrating equipment of the three suppliers has been the topic of numerous meetings and negotiations.

To understand the difference in capacity between TAT-8 and TAT-7 (the seventh metallic cable to cross the Atlantic) is to understand why the last transoceanic coaxial cable has been built and why the first such fiber optic system is coming into being. TAT-7 had the capacity for carrying 5000 voice circuits; TAT-8 has the potential to carry 37,800.

When installation of TAT-8 began in 1987, a sea plow towed by a cable-laying ship must furrow the continental shelf. The plow cuts a slot two feet deep in the sea floor; cables and repeaters then pass through the hollow plowshare and are deposited in the bottom of the trench. (See Figure 4.)

---

[3]"TAT-8 Hopefuls Submit Bids; AT&T Claims " 'We're Number One," *Fiber Optics News*, May 27, 1983.

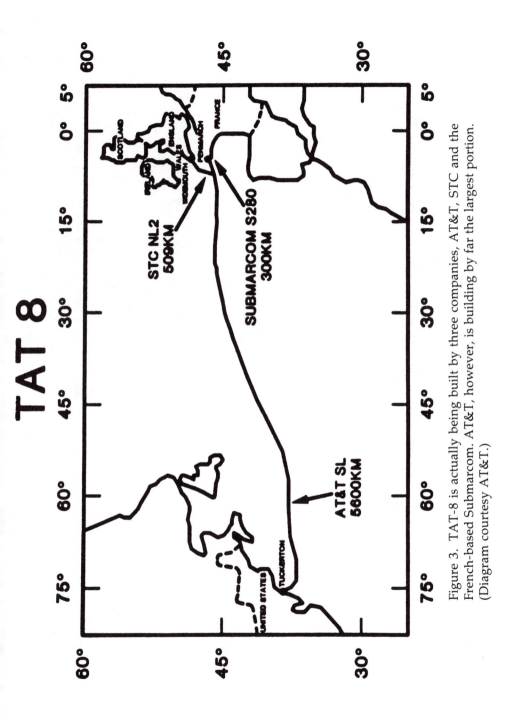

Figure 3. TAT-8 is actually being built by three companies, AT&T, STC and the French-based Submarcom. AT&T, however, is building by far the largest portion. (Diagram courtesy AT&T.)

Figure 4. In areas where fishing trawlers constitute a real threat, AT&T employs its sea plow, towed by a ship, to bury cable. (Photo courtesy AT&T.)

It might be tempting to cite TAT-8 as an example of good old American ingenuity winning out. But, here again, there is a Japanese connection. Hitachi is providing the semiconductor laser chips for AT&T's portion of TAT-8.[4]

Four communications services—geared to various sized business users—are expected to be offered over TAT-8, at least initially. Included will be digital communications at 56 and 64 Kbps and at 1.544 and 2.048 Mbps.

[4]"Hitachi to Provide AT&T With Semiconductor Laser Chips for TAT-8 Terrestrial Systems," *Fiber Optics News,* March 2, 1984.

TAT-8 is giving birth to associated new products, including Digital Circuit Multiplication systems and a hybrid digital multiplexer.

TAT-8 is rapidly taking the gleam off the rose of the satellite industry. International satellite provider Intelsat was predicted to lose $500 million in business as the result of TAT-8, according to Hughes Aircraft Company's Bruno Miglio. Perhaps recognizing this, Intelsat U.S. partner Comsat fought bitterly to have the FCC delay the cable landing license for TAT-8, arguing that the additional circuit demand did not warrant the 1988 cutover of TAT-8.

In harshly rejecting the Comsat request, and ordering the TAT-8 cable landing license, the Federal Communications Commission explained that the "introduction of TAT-8 will mean the introduction of digital fiber optic technology with its attendant service benefits." Fiber optics, the commission said, will also provide the countries so linked with the capability of connecting their own digital links to this all-digital network. TAT-8 itself "will spur the development of domestic digital systems." The fact that TAT-8 is there could also stimulate demand for services that heretofore would not even have been contemplated. And, the FCC said, "TAT-8's fiber optic technology can transmit video and other broadband services now available efficiently only over satellite."[5]

Because TAT-8, with the large amount of capacity it can afford, would mean that a reasonable amount of traffic would go back to cable and away from satellite, AT&T tried to get the FCC to stop the balanced loading formula, which stipulated that 60% of trans-Atlantic traffic had to be carried over satellite. The FCC decreed that such a shift of 2% annually can occur through 1988, with decisions about balancing after that to be made later. The FCC has given indications, however, that as part of its deregulatory push it would like to get out of the business altogether.

[5]FCC Granting of TAT-8 Cable Landing License (File No. S-C-L-84-001). Adopted May 24, 1984, released June 13, 1984.

As part of its deregulatory mode, the commission has stipulated that private networks can also come into being without the scrutiny that public networks have been subjected to. This was true for international networks, as well. The commission's hope was to create a more competitive marketplace.

The satellite industry attempted from the beginning to fight the use of fiber optics for undersea applications. Each early application by AT&T and other U.S. international record carriers for an international fiber optic network was met with hostile, at times shrillish, charges by Comsat that the network would provide too much capacity, had not been tested, would cost too much, etc.

It is no wonder the satellite industry has come out swinging. After making magnificent gains as a glamour technology, satellite is being replaced by a newer glamour technology—and stands to lose billions of dollars as the result. One magazine editorial entitled "Turbulent Times for Satellites" summed up the problems for the satellite industry in two words: optical fiber.[6] *The Wall Street Journal* noted that the satellite industry even stands to lose the traffic it now has as carriers switch off satellite and onto fiber.[7]

Some in the satellite industry have noted that it might make more sense to try to accept the advent of fiber optics and make the most of it. This is happening in some instances, and there are projects, such as teleports, where the two technologies can work in harmony.

It is becoming increasingly evident that those who believe only in satellite to the exclusion of fiber optics are on a kamikaze mission. Having lost the argument to block TAT-8, Comsat has since reconsidered and hired on to lease capacity on the system. (The Comsat group responsible for signing with TAT-8, Comsat

[6]"Turbulent Times for Satellites," *Communications Systems Worldwide*, February 1986.

[7]"Future of U.S. Firms in Lofting Satellites is Far From Assured," *The Wall Street Journal*, September 30, 1986.

International Communications, Inc., has since been purchased by Contel.) Comsat also uses more than 1000 miles of fiber optic capacity as part of its terrestrial networking. The American Satellite Company in Rockville, Maryland, leases a large amount of fiber optic capacity from Lightnet and depends on fiber as its primary transmission medium between New York and Chicago.

Comsat International Laboratories Director John Evans has acknowledged that it will be very difficult for the satellite industry to compete when the world is largely fibered. In a sharp departure from positions it has held in the past, Evans noted in 1986 that Comsat is "not wedded" to satellite technology.

"The competitive situation for fiber optics will be somewhat ameliorated by the failures of satellite launch vehicles," says KMI's Holtzman.

TAT-8, in fact, is only one component of what Holtzman calculates will be a $5.4 billion market by 1995. This involves 100,000 cable km or 600,000 fiber km. This, in fact, could reach an additional 65,600 cable km and 394,000 fiber km. Combined planned and proposed systems total 1 million fiber km, Holtzman figures.

While this sounds as though it may create a glut of capacity, Holtzman analogizes the situation to that of the development of the U.S. fiber optic long-haul market. "With a few exceptions, most of the announced U.S. long-haul systems are being built and are proving viable," he said. "The increasingly competitive nature of the international communications market will allow most of the planned undersea systems to succeed financially."

Two private-line ventures that would compete with TAT-8 arose, Market-Link, the previously mentioned Tel-Optik and Cable & Wireless effort, and Submarine Lightwave Cable. What the proposed $600 million Market-Link has going for it is the support of Cable & Wireless, which has charted its own path of establishing a "global digital highway." Two fiber optic cables, with parallel routes, are being offered, PTAT-1 (privately owned trans-Atlantic submarine cable) and PTAT-2. The first is planned for operation by 1989 and the second by 1992. The network was originally announced as having the same technology TAT-8 would use and is seen as being so similar to TAT-8 in scope that one wag has declared it "Copy TAT."

The other proposed network, the Trans-Atlantic Video submarine cable system (TAV-1), was offered by a group known as Submarine Lightwave Cable Company. Its aim was to provide video services between the United States and Europe.

Both were to eventually receive cable landing licenses from the FCC, although the Commission was more reluctant to act on the partnerless SLC bid. Of note is the fact that AT&T supported FCC approval of both ventures, standing by its argument for a deregulated international marketplace, even though each could provide competition.

The Market-Link venture has taken on an interesting wrinkle with NYNEX hoping to buy out Tel-Optik's shares. The Regional Holding Companies under Judge Greene's modified final judgment are prohibited from providing exchange service outside of their designated areas. As such, NYNEX must obtain a waiver from Judge Greene to complete the transaction. To this point, Greene has allowed NYNEX and Tel-Optik to go ahead with a plan placing Tel-Optik's shares into a trust run by Tel-Optik officials. This is despite an objection by AT&T, who seemingly underwent an attitude change when former partner and potential competitor NYNEX came into the picture. In addition to the $10 million it would pay for the Tel-Optik shares, NYNEX would also assume Tel-Optik's debt, a figure between $150 million and $200 million, according to NYNEX. That represents less than two percent of the value of NYNEX's outstanding stock. (See Figure 5.)

One reason AT&T may not be that concerned with the competition is that it intends to follow up with a second trans-Atlantic fiber optic network, TAT-9, in 1991. TAT-9 originally was not scheduled to come on board so soon—the rule of thumb is a new TAT cable every five years—but competition has acted as a driving force.

Despite the small interval, TAT-9 will offer significantly

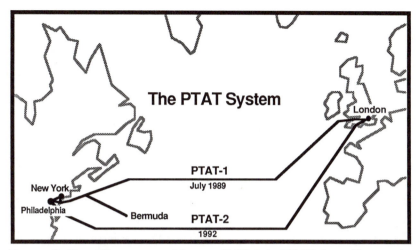

Figure 5. Providing competition to TAT-8 is PTAT-1, a fiber optic cable being constructed by Cable & Wireless and Tel-Optik. NYNEX is attempting to purchase Tel-Optik's ownership in the deal. (Diagram courtesy Tel-Optik.)

advanced technology over TAT-8. The $400 million network will operate at 565 Mbps over two fiber pairs, providing the equivalent of 80,000 voice grade circuits.

TAT-9 will include Canada and Spain as well as the United States, France, and the United Kingdom. Spain and Canada were also in on TAT-8 discussions, even though they were not included. A sophisticated branching unit will route traffic to three locations on the European side and two in North America.

The TAT-9 consortium does not have to fear objections from Comsat when the group files with the FCC. Perhaps learning that "if you can't lick 'em, join 'em," Comsat is one of the partners in the TAT-9 venture.

Plans are also afoot to build a Mediterranean network with Palermo, Italy, to serve as the central node, with fiber routes joining Palermo with Tel Aviv, Israel, Marmaris, Turkey, and Lechaina, Greece. A separate route may link Palermo with Spain and several Arab countries.

In the Pacific, AT&T intends to build a fiber optic network

connecting California with Hawaii (HAW-4). This will be connected to a fiber optic network that (via a mid-ocean branching unit) will extend to Guam and Japan (TPC-3). TPC-3, in turn, will connect to a fiber optic network connecting Guam with the Philippines (GP-2) and with a network coming up from Australia.

As planned, AT&T will build all of HAW-4, some 2500 miles, and GP-2. AT&T will also build TPC-3 to a point just east of the branching unit to Japan (2400 miles), where KDD will take over and build the remaining 1500 miles to Japan and another segment to Guam. The 7200-nautical-mile HAW-4/TPC-3 will cost $593 million. The 1500-mile network joining Guam and the Philippine island of Luzon is predicted to cost $107 million.

Construction work at the shore stations could begin in 1987. The deep-sea portion of HAW-4/TPC-3 could be installed during the summer and fall of 1988, while the deep water work on GP-2 could occur in early 1989.

Providing competition to HAW-4/TPC-3 is Pacific Telecom Cable Inc. This cable is 50% owned by partners at the US end, and fifty percent owned by partners at the Japanese end. Partners in Japan include C. Itoh, Toyota, several major Japanese banks, and potentially Regional Holding Company Pacific Telesis. Not surprisingly, Cable & Wireless owns 20% at each end. The reason this is not surprising is because Cable & Wireless Chairman Sir Eric Sharp has let it be known that he intends to wire the world's major financial centers with fiber optics. In an article entitled, "Can Sir Eric Sharp Ring The Earth With Glass Cable?" *Business Week* said Sharp intends to have New York, Hong Kong, and Tokyo connected by using a $3 billion digital network. One would also have to include London if PTAT-1 meets its deadline. Not surprisingly, old Cable & Wireless crony STC has received the major contract to build PTAT-1.[8]

---

[8]"Can Sir Eric Sharp Ring The Earth With Glass Cable?" *Business Week*, November 3, 1986.

The $500 million Pacific Telecom cable route is expected to link the states of Washington and Alaska with Japan. The network would feature three fiber pairs operating at 280 Mbps or higher. The Pacific Telecom Cable route has become a bit of a political hot potato, as the Japanese did consider not allowing a KDD competitor without large majority ownership by Japanese firms. Prime Minister Margaret Thatcher and the U.S. Congress have complained bitterly about that course of action. More recent developments indicate the Japanese will allow the Pacific Telecom Cable route to go forth, as well as a separate system owned mainly by Japanese companies, according to knowledgeable sources.

AT&T and four other telecommunications administrations have also planned a fiber optic network connecting Hong Kong, Japan, and South Korea by 1990. The H-J-K cable is slated to have a total length of 2460 nautical miles and cost $200 million. The network will interconnect with TPC-3. Other carriers include KDD, Cable & Wireless, the Korean Telecommunications Authority, and Telecom Singapore.

The Australian Overseas Telecommunications Commission (OTC) also has aggressive plans to reach out with fiber. Australian Prime Minister Bob Hawke at the end of 1986 announced a solicitation of $1 billion (Australian) for that purpose. Early plans call for Australia to be joined with a northern island of New Zealand by 1991. The cable will reach 2500 km. Interconnection with North America and the Far East over fiber is also anticipated.

Australia's economic future depends on the establishment of communications networks, according to George Maltby, managing director of the OTC. And that will depend largely on the growth of fiber optics, which will replace satellite as the major communication system serving the area.

AT&T also plans to construct a Caribbean fiber optic network connecting Florida with a hub in Jamaica and including branching units for the Dominican Republic and Haiti. An extension from Jamaica to Colombia, South America, is also planned. As mentioned, AT&T is also building a "domestic" link to Puerto Rico from the mainland United States.

For fiber optics, rewiring the oceans of the world represents a unique and exciting challenge. Dr. Kao believes there will be a push for the repeaterless spans mentioned at the beginning of the chapter. Bell Laboratories scientists have joked that TAT-15 might be the one.[9] This author has speculated that March 10, 2026, might be the day for AT&T to inaugurate the first repeaterless transoceanic span. It would be 150 years to the day from Alexander Graham Bell's historic message: "Mr. Watson, come here, I want you."[10]

Certainly the economic incentives to design such a repeaterless network are there. First of all, the cost of repeaters is significant and companies are moving to eliminate them. (There are scheduled to be approximately half as many repeaters on TAT-9 as there are on TAT-8.) Without having to worry about repeaters, engineers could build in even greater levels of reliability for these undersea networks.

If such an unrepeated pulse could be sent thousands of miles, it is a cinch that it would not be using the basic building blocks that have brought fiber optics this far. [Other things will change as well; scientists are working on a computer system that will automatically translate language if people speaking two languages are on the same telephone line.]

The key ingredient may be fluorine fiber, whose theoretical losses are much lower than those of silica. Whatever the answer, the smart money is betting that—using fiber optics—scientists will find a way to solve yet another fiber technological challenge.

[9]"Repeaterless Transoceanic Span May Be Selling Feature of TAT-15," *Fiber Optics News*, March 24, 1986.

[10]"Fiber Optics Leap Into the 21st Century," by C. David Chaffee, *Optics News*, September/October 1983.

Chapter 13

A World of Applications

PRESIDENT RONALD REAGAN has personally experienced the advantages of fiber optics in two ways that have nothing to do with voice or data communications. Optical fibers have helped to locate polyps in his colon that have then been removed. The President and Mrs. Reagan have enjoyed a fiber optic ornament supplied by Corning Glass Works, which has adorned their Christmas tree.

Potential applications of fiber optics seem limited only by human imagination. Optical fiber has illuminated wedding dresses in Japan for nocturnal ceremonies. Optical fiber cable has snaked through the rubble of a Mexican earthquake to provide images of who or what might need to be saved. It is also being used to help detect earthquakes before they happen.

Decades ago it was used to light hood ornaments on fancy automobiles. Today, automotive experts are incorporating it in a variety of roles.

Besides realizing that fiber optics can make their craft lighter and less immune to interference, aviators are finding that fiber optics can play an important role in air traffic control.

The sky is really not the limit for the use of fiber, as foresighted engineers have already developed a number of ways that fiber can enhance—and benefit from—space exploration.

Medical Applications

Fiber optics has been used in medicine longer than it has in communications. Bundling thousands of fibers together to make up what was considered a small area, doctors have been able to obtain good images of peptic ulcers, cancerous polyps, tumors, and other unwanted tissue. The technology is still being used for imaging, but it is also being implemented in a host of other applications, including the treatment of cancer and heart disease. Because techniques that use optical fiber and lasers can zap or destroy unwanted tissue by strong beams of light, photonics may well revolutionize surgery as we know it.

In fact, the use of fiber optics in medicine is such a growing phenomenon that Dr. Abraham Katzir predicts it will continue to assume a larger slice of the overall fiber optics pie. A professor at Tel Aviv University, Boston University, and MIT, Katzir believes medicine could eventually account for 50% of the entire fiber optics marketplace.[1]

Microsurgical techniques in which fiber optics and lasers are used can be applied when "you have a small thing in the body that interferes with the whole system," says Katzir. This can include a blood clot or plaque. There are millions of these operations performed every year which could use fiber optics, according to Katzir.

There are three general categories into which the use of fiber optics in medicine fall, including imaging, diagnostics, and therapy. The high-resolution images are fed to television screens by fiber optic endoscopes, known as fiberscopes. Fiberscopes can be inserted in the body through an orifice or by cutting the skin.

Fiberscopes can be used to carry images from areas such as

---

[1]"Medical Applications Seen Climbing to 50% of Fiber Optics Market in Next Decade," *Fiber Optics News*, September 29, 1986.

the colon, gastrointestinal tract, and pulmonary system. (The President's polyps were found in the colon.) Fiberscopes can also be used to examine the urology tract. Indeed, almost every discipline of medicine, with the exception of dermatology, uses some type of endoscope, according to Ann Pilson, product manager for Reichert Fiber Optics, a fiberscope supplier.

Fiberscopes use a lens at one end of the fiber, which is inserted in the patient; another lens is placed at the viewing end of the fiber. The endoscope uses two fiber bundles, one to carry the image from the body and one to illuminate the area being viewed. There are two basic types of endoscopes: 1) rigid endoscopes to examine joints such as knees and ankles, and 2) flexible endoscopes, which can study the stomach and other areas. Flexible endoscopes almost always use fiber.

Before fiber optics technology was available, it was difficult to examine certain parts of the large intestine, such as the secum (the section closest to the small intestine), according to Dr. Stanley Benjamin. One type of fiberscope, the colonoscope, sees all the way to the end of the large intestine, a task difficult for more conventional technology. "It is a testimony to fiber optic technology" that more thorough examinations are possible, according to Benjamin.

More streamlined fiberscopes, known as imaging cathoders, house only 1000 to 2000 fibers, according to Katzir. While these provide lower resolution, they are more affordable. Thin endoscopes can now be located in the heart, in situ, to observe valves or other important areas.

The transmission of holographic images provides an extra dimension to imaging activities. Holography would allow doctors to even more precisely locate a cancerous tumor, for example, by gaining a three-dimensional image. The need for an exact location is important when tumor surgery is performed with a high-powered laser. (See Figure 1.)

Fiber optic sensors are employed in medicine to measure blood flow, oxygen, pressure, and temperature. For example, a fiber optic sensor has been used to measure the temperature pro-

Figure 1. One potential exciting application of fiber optics is working with holograms. Doctors, for example, can gain a better understanding of objects in the body if the data is communicated three dimensionally. Not surprisingly, AT&T Bell Laboratories has explored applications in this area. Here, T. Dixon Dudderar holds the end of the coherent bundle of optical fibers, which transmits holographic data from the circuit board under observation. The end of the fiber optic strand carrying the illumination beam from the laser can be seen at right foreground. (Photo courtesy AT&T Bell Laboratories.)

duced by a laser beam in order to destroy a cancerous tumor. At 42 degrees centigrade, the tumor would die but the surrounding tissue would not be significantly damaged. The sensor is inserted in the tumor to reveal when the temperature has been met. Fiber optic sensors in medicine are often attached directly to optical fiber.

Many of the fibers being developed are significantly different from the silica based fibers used for communications. Rather than carrying relatively low-powered signals continuously, the fiber must be able to handle power levels of 1 to 5 watts for short durations (10–100 nanoseconds). The larger the obstruction is, the larger the amount of power required to destroy it. For example, power levels of 100 watts have been reached outside the body; such lasers could conceivably shatter gallstones inside the body.

Of the three most promising methods of using lasers and optical fiber to clean up blood vessels by removing plaque (laser angioplasty) being explored in the mid-1980s, none uses the silica based fiber that is popular in communications. These processes include zirconium fluoride fibers using a YAG laser operating at 2.9 microns; silver halide fibers powered by a carbon dioxide laser; and quartz fibers coupled with an excimer laser.

Popular applications of fiber by the mid-1980s included the use of neodymium YAG lasers for coagulating blood and using fiber optics in opthalmology to replace mirrors used to provide an image of the inner eye. Fiber optics are also used in ear operations.

Dentists have also discovered the advantages of fiber optics and are using the technology for illumination and as a substitute for x-rays. In the 1960s, fiber optics was used for illumination purposes only. It was not until the 1970s that fiber was used for diagnostics.

Fiber optic transillumination (FOTI) involves a bright light carried by a fiber optic probe to illuminate a patient's teeth. The probe is held behind the tooth so the light shines through, similar to candling an egg. Areas in the tooth that contain cavities show up as dark spots.

FOTI is demonstrating success rates approaching those of x-rays, providing a valid substitute for pregnant women who cannot be exposed to radiation. Children who cannot sit still or who have problems biting down on the cumbersome bitewings necessary in x-ray situations are also candidates for its use.

Entertainment

When Walt Disney Productions' Experimental Prototype Community of Tomorrow (EPCOT) opened in October 1982, it was fiber optics that helped to keep the customers satisfied. EPCOT's world key information service (WKIS) consisted of 29 video screens connected to a central computer via multi-mode fiber. The interactive video system becomes activated when infrared beams penetrating the screen sense a visitor's touch. The visitor is provided with information about EPCOT's features, including motion picture previews of activities, restaurant menus, motel vacancies, and a map showing the easiest way to get to another exhibit. Information is presented in English and Spanish.

WKIS has a built-in alarm and maintenance system. When it first came into commercial use, it used four laser disks in the unit to provide video information when needed. Each disk stored 54,000 television frames in digital form. Communications among the 20 pavilions is also carried by fiber optics.

What was then Western Electric, instrumental in WKIS, cited two main reasons for using fiber. For one thing, Orlando, Florida, EPCOT's home, has the second largest incidence of thunderstorms in the nation, according to Peter Krawarik. "The second advantage is in quality," he said. "The information-carrying capacity of optical fibers permits broadcast studio quality on the TV screens, which is not easily available over traditional copper cables."

The ABC Radio Network has transmitted audio signals between Washington, D.C., and New York via optical fiber cable, and the other networks have also used it.

Major New York film studio Kaufman-Astoria, a customer of Teleport Communications, is now wired with fiber optics. Due to its interconnection with satellites on Staten Island, the teleport has the capability of sending signals from any satellite in the world.

Car stereo systems using fiber optics have also appeared. An early unit introduced by Alpine Electronics of America allows the driver to select up to six cassette tapes for continuous play. The cassette changer is located in the trunk of the car with programming commands carried by optical fiber from the driver's computerized control console. Only programming commands are carried over the fiber.

The Japanese Tomita-ai Shoji Company has developed a bridal gown, Merveille, which uses optical fiber as yarn. Some 10,000 pieces of optical fiber are embedded in the costume, which was listed at a cost of $2,512. Expected to follow are other types of dresses and tuxedos sporting embedded optical fibers.[2]

While the market for using fiber optics with cable television has not evolved as rapidly as some had thought it would, it still represents the potential of a major market.

Working with Plessey Stromberg-Carlson, Northwestern Bell has developed electronics allowing for the transmission of three television signals, each assuming a capacity of 45 Mbps. How soon this and other examples of digital video come to market, however, depends on economic factors, according to Jules Bellisio, manager of Bell Communications Research's Signal Processing Research Division.

Bellisio believes there will be a "tug-of-war" between the price of optical fiber cable and the development and marketing of equipment that can provide entertainment-quality channels at lower bit rates per channel, such as 20 Mbps. Efforts to squeeze more video into the fiber are ongoing.

It is no surprise that a telco such as Northwestern Bell is involved. Cable television guru Irving Kahn believes the telcos will play an important role in the bringing of fiber to cable television. One possibility is the use of fiber to transmit high-definition television to the home. Prevention from fire and burglary through sensors are other possibilities, as are a host of other services.

Energy/Environment

The U.S. Military Academy at West Point is using fiber optics for its energy management and control system. A central computer is tied into remote terminal units dispersed among 40 to

---

[2]"Wired Bridal Gown Glows on That Very Special Day," *The Asian Wall Street Journal*, September 29, 1986.

50 buildings on campus. The terminals, which include temperature sensors, will send data to the computer via optical fiber links, monitoring temperature and energy use, and controlling heat and ventilation.

Williams Electric Company received more than $6 million to install the system. This and other types of energy management systems are "going to go more and more to fiber optics," according to Aubrey Gooner, Williams' project manager on the job. While such systems in the past were installed using existing telephone lines, those copper based systems have been found to be susceptible to electrical storms and noise.

Fiber optics is playing an important role in the search for oil. It is being attached to logging tools going down a well or to an ocean's bottom to feed up important data. The design of one such system by Optelecom and Chevron can carry up to 300 times more information than previous logging tools were able to handle.

Working with Artel, Chevron has also installed a fiber optic link at the headquarters of Chevron Geosciences in San Ramon, California. The fiber system allows for the transfer of video images among various workstations located in separate buildings on the complex. The images represent a seismic photograph of areas where oil is suspected to be buried beneath the surface.

A third application pertaining to the oil industry involves a fiber optic process control system, which will aid in monitoring oil production in Alaska's north slope oil field. The system is to be installed at Sohio's Endicott field production facilities located off shore. It will monitor the oil recovery stage of production at the facility.

The system, which is being designed by Advanced Fiberoptics Corporation, will provide data concerning the status of valves and other oil-carrying conduits.

The advantages of fiber optics have long been recognized by the commercial nuclear energy industry. Artel led the way in developing a fiber optic transmission system that can continuously produce multicolor video computer graphics, thus helping to ensure the safe operation of a nuclear plant. This monitoring equipment has the advantage of providing data from highly vulnerable areas, without the threat of interference rendering the monitoring equipment useless.

To guard against fire and heat in a nuclear plant, suppliers such as Chromatic Technologies have developed flame- and radiation-resistant optical fiber cables that comply with Nuclear Regulatory Commission standards. The cables can be used for video monitoring, remote process control, and data communications.

Optelecom and other companies are also developing ways in which to send robots into potentially contaminated nuclear sites to aid in rescue and cleanup operations. Fiber's immunity to interference and its ability to transmit large amounts of information give it intrinsic advantages over other types of links. These robots could also be used for fighting fires and dealing with unexploded ordnance.

In addition to these applications, protection from intrusion and sabotage in restricted areas is being monitored by patterns made of optical fiber incorporated into seals. Used in some nuclear plants, altered patterns reveal intrusion on-site, unlike more conventional units.

By lowering a pair of optical fibers connected to conventional fluorometers over the side of a ship, the Canadian Department of Fisheries and Oceans is finding out more about fish-feeding habits and ocean ecology. By measuring an element of photoplankton to depths of 100 meters, scientists are able to gather more knowledge about a particular area.

Sensors

Fiber optic sensors are an outgrowth of fiber optic communications. The annoying losses that scientists discovered when optical fiber was subjected to pressure or stress for use in communications systems were found to be constant and measurable. These specific sensitivities could be used to measure tempera-

ture, pressure, acidity, acceleration, and other characteristics
when placed in the desired environment. (See Figure 2.)

Basic configuration of a fiber optic sensor system consists of a
light source, the fiber connecting the light source to the sensor,
and a detector, which determines if changes have taken place.
Corresponding to the development of the fiber optic commmun-
ications market, LEDs and multi-mode fibers reached promi-
nence first, followed by lasers and single-mode fiber.

Fiber optic sensors have certain inherent characteristics that
make them more attractive than conventional sensors. For one
thing, they are resistant to corrosion. They are also immune from
interference and are lightweight.

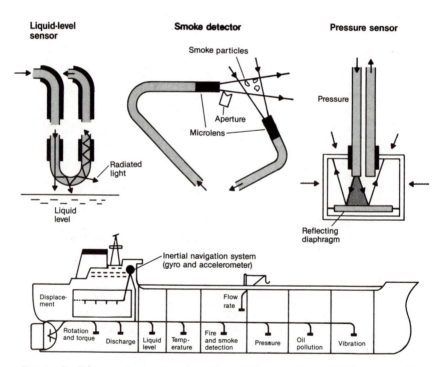

Figure 2. Fiber optic sensors can be used in a variety of different applications, including measuring liquids, smoke, and pressure. The vessel
pictured at the bottom uses a variety of different sensors. (Diagram
courtesy *IEEE Spectrum*.)

While prognosticators vary in their estimates of how rapidly the market for fiber optic sensors will grow, each believes it is a high growth area that will accelerate from the tens of millions of dollars to the hundreds of millions of dollars from the mid-1980s to the early 1990s.

Optical fiber sensors were first commercially used in measuring instruments and controls where sensitivity, resistance to hostile environments, and compactness were essential. The first commercial sensors measured temperature, pressure, acceleration, flow of liquids and gasses, and the level of a liquid in a container.[3]

One application has been to monitor oil spills on ships. The sensors are sensitive enough to differentiate between marine oil pollutants and solid pollutants. Such devices were sailing on more than 1000 vessels by 1986.

The National Bureau of Standards announced in 1982 that it was developing an optical fiber thermometer that could measure temperature to 2000 degrees centigrade, a 500 degree increase over existing standards at the time. The thermometer's developer, Ray Dils, predicted the thermometer would be able to measure temperatures to an uncertainty of less than 0.01%.

Dils left NBS in 1983 to start Accufiber Inc. in Vancouver, Washington. The company's first product was the optical fiber thermometer, which Dils priced at $9500. Initial applications were for measuring gas turbine temperatures. Dils has also developed it for the automotive market and believes it could improve engine efficiency. He received a patent for the thermometer in 1986.

The NBS continued to use the thermometer after Dils left. The agency hopes to develop a more accurate temperature scale from 640 to 1064 degrees centigrade. NBS is also attempting to determine accurate temperature values for the freezing points of silver and gold relative to that of aluminum.

---

[3]"Optical Fiber Sensors Challenge the Competition," by Thomas Giallorenzi, Joseph Bucaro, Anthony Dandridge, and James Cole, *IEEE Spectrum*, September 1986.

Optical Technologies in Herndon, Virginia, has designed and built the world's first fiber optic seismometer—an instrument the United States will use to monitor Soviet nuclear explosions. The device has also been used to predict accurately the impact of earthquakes.

An optical fiber probe developed by Britain's Meat Research Institute has been used to stress-test meat carcasses. The testing is important because meat is believed to be more tender when it undergoes less stress prior to and during an animal's slaughter.

Fiber optic hydrophones—used to detect underwater sound—have already been tested at sea, according to officials at the Naval Research Laboratory, which is actively developing them.[4]

While comprising a small market in the early 1980s, fiber optic gyroscopes are expected to garner an annual market of upwards of $120 million by 1994. An early fiber optic gyroscope developed by McDonnell Douglas for logging data on oil wells can operate to depths of 600 meters at temperatures from 0 to 125 degrees centigrade.[5]

Transportation

The Ford, Chevy, Toyota, or Chrysler you may be driving probably uses limited amounts of optical fiber, and the car you trade that in for down the road is likely to incorporate more significant amounts.

While hood ornament lighting belongs to a bygone era, fiber optics in the mid-1980s was being used for backlighting instrument panels and direct lamp monitoring.

In one of its models, Nissan Motor Corporation employed a link allowing a passenger in the rear of the automobile to control the air conditioner and car radio. In another, an optical system located on the steering column transfers a light beam through the air to a sensor located on the car's roof. The beam carries

[4]Ibid.
[5]Ibid.

commands for the automobile's radio and automatic speed control.

The first features in automobiles to reach widespread service by fiber will be in the convenience accessories area, including window controls, air conditioning, etc., according to Frank Dixon, president of Electronicast.

The first area to be integrated with fiber will probably be the wiring harness on the car's steering column. The current arrangement is bulky, with a pair of wires for each instrument. To remedy the problem, there may well be a fiber power bus connecting all instruments and equipment. Controls on the dashboard will be push-button operated. Power will come from the automobile battery.

There are already indications that suppliers are gearing up to satisfy the automobile harnessing marketplace. Alcoa-Fujikura, whose original intent was to provide optical power grounding wire to utilities, has purchased PEP Industries, a manufacturer of wire and cable harnesses for automobiles and trucks. The direct aim of the acquisition is to incorporate fiber optics into automobiles.

In addition to being used in automobiles, fiber is also helping to direct cars on the road. Traffic signs you may remember seeing on the Delaware Memorial Bridge or other bridges that are used for lane control and use red X's and green arrows may use optical fiber illuminated by halogen lamps. Colors can be changed by using filters. Word messages, such as "Accident Ahead," can also be illuminated by using fiber bundles. Another application includes traffic monitoring by jurisdictional bodies to monitor traffic flow and keep an eye out for accidents.

Boeing Commercial Airplane Company has adapted a fiber optic interconnect system. ITT's Cannon Electric Division was the first to develop such a system.

This is separate from the U.S. military's commitment, which has been exploring "Fly By Light" systems since the 1970s. By 1985, the F-16 had 3 fiber optic data buses on board; the F-18 had 1, the space shuttle had 28, and the F-15 was being refitted for that purpose.

The Federal Aviation Administration is now testing a prototype fiber optic system at the Meadows Field airport in Bakersfield, California, to provide air traffic control.

"Radio interference at an airport is a real problem, and it's getting more and more difficult to get frequencies," said Neal Blake, the FAA deputy associate administrator for engineering. As the result, the FAA is looking to alternate systems such as fiber optics. Government solicitations indicate these systems may be widespread in the near future.

Delta Airlines has installed an 8-route-mile ring network architecture at its facility at Hartsfield Atlanta International Airport. Delta was hoping to increase its capacity without using additional cable. The system is fault-tolerant; it can reroute signals automatically to a standby circuit in the event of equipment failure, node outage, or cable cut.

Outer Space Applications

The Space Station expected to be hurtling gently through space in the coming years will be adorned with large arrays of fiber optic sensors seeking out errant meteors and gathering information for the control center. That is, if work being done by McDonnell Douglas engineers comes to fruition.

Proponents of fiber optics have been trying at least since 1982 to get NASA specifically and the U.S. government generally interested in installing fiber optics on the Space Station.[6] Certainly one application is in including a fiber optic local area network inside the vehicle, and NASA Johnson Space Center officials have expressed keen interest in this. Such a network would be used for data processing, distribution, evaluation, command generation, and storage.

"Integrated and fiber optics are particularly suitable for space payload by virtue of their advantages, especially weight and

---

[6]"Optical Fiber Backers Begin Drive to Include Technology in Space Station," *Fiber/Laser News*, November 12, 1982.

power consumption," according to Talal Findakly and Fred Leonberger of the United Technologies Research Center. "Such markets may open sooner than anticipated given continued interest in the Strategic Defense Initiative and other space-related programs."

To show its interest in providing such a LAN, RCA has rolled a model of its 200 Mbps fiber optic LAN into the Johnson Space Center Electronics Systems Test Laboratory. The 200 Mbps LAN could be upgraded as the data requirements of the Space Station change during its 30-year lifetime.

NASA's Deep Space Communication Complex and the Naval Research Laboratory have already begun testing fiber optics to gain an indication as to how it would perform in outer space. Optical fiber has also been sent up in the Space Shuttle to see how it would perform. The scope of the NRL work is to irradiate fiber in low doses in a way similar to exposure in space.

Outer space has the potential for processing glass used to manufacture optical fiber to purity levels that are just not possible on earth. This is one suggested application aboard the Industrial Space Facility, which could be launched by as early as 1990. Theoretical levels to this point carry a TBD (to be determined) tag, with the potential for extremely low-loss glass, resulting in much lower attenuation levels, a real possibility.

Chapter 14

Networks of Tomorrow

T HE REWIRING OF America will be complete only when optical fiber joins the office buildings and homes in the United States with the supertrunks that are already spread throughout the land.

The wiring of America's offices with optical fiber has already begun en masse, kicked off by the rewiring of Miami's financial district in 1984.

Two years later, it was becoming a common occurrence. "Every day now some building gets wired directly with fiber into the network," according to Robert Lucky, who was interviewed in late 1986. Lucky is executive director of Bell Laboratories' communications sciences division.

"As new buildings are built, their developers want the latest communications technology and many insist on fiber," explained Lucky. "Business developers are interested in strategic communications because people are more interested in being in a place with lots of communications than they are in being in a beautiful place."

Lucky envisions breakthroughs in the late 1980s in the ability of local networks to carry information. Streams reaching 100 billion bits per second could be used in these LANs.

"These rates are so huge that frankly we haven't any idea what the computer community would do with them," he says. "The computer community doesn't know either, but I think they would find ways to use this, I think we all would." Lucky sees one end result as our "wasting a lot of bits to do things easier."

When it comes to bringing fiber to the home as compared to the office, however, Lucky rings the caution bell. AT&T has come to the realization that the way it will bring fiber to the home is through voice telephone, and that could take a while. The cost of using the millions of components necessary to bring fiber to the home was prohibitive, at least by 1986, 5 to 10 times what it would have to be to get the job done economically. "We have got to get cheap stuff, that's the name of the game," he says.

Bell Labs and BellCore are feverishly working to bring the cost of these components down to the level where they could be incorporated into the home, as are scientists from numerous other groups.

To further explore this area, OFC/IOOC '87 held a special panel the evening of January 20 entitled "Fibers to the Home and Business—Opportunities, Challenges and Timing," chaired by BellCore's Stewart Personick. While none of the panelists disputed that fiber to the home will eventually become a reality, various practical and technical concerns need to be addressed before that day comes. [In 1987, OFC was coupled with the Sixth International Conference on Integrated Optics and Optical Fiber Communication.]

Types of services to be offered over such systems—and who would want them—represent one problem. Panel members challenged the audience to find new innovative services that could be offered over the fiber that could help generate revenues to pay for the systems. For example, only a modest percentage of those with access to cable television even subscribe to it, said Matthew Miller, Vice President of Science and Technology at Viacom International.

Yet the problem of matching what the subscribers want with

what is offered could well be just a process of trial and error. The French Minitel project, in which customers are provided with terminals instead of telephone books, has received good response among users, who access weather reports, stock quotes, and other specialized services over a copper network that joins approximately 3000 users. "The most popular is the dating service," reports Corning Senior Vice President and General Manager Jan Suwinski.

Bringing fiber to the home could also aid in the growing use of telecommuting, the phenomenon through which one in three workers is expected to operate out of his or her home by the year 2000, according to Richard Teresi, the keynote speaker at OFC/IOOC '87. Teresi is editor-at-large of *Omni Magazine*.

One way of bringing fiber to the home that is gaining increasing attention is for the utilities to do it. This requires public service commission approval and will obviously have to result in attitude adjustment among state board members, many of whom are not quick to accept change. Certainly if the economics are there, however, that represents a viable alternative. What is not considered an alternative, however, is subsidizing these kinds of networks before they cost in economically, said panel member Louise McCarren, of the Vermont Public Service Commission.

Corning envisions the evolution of communications, entertainment, and energy management in the home as a gradually evolving scenario. Homes in the recent past deployed limited communications, entertainment, and security processes that were disparate. Home communications today is more sophisticated, with the incorporation of additional communication facilities, security, and entertainment. Still, however, these various services remain segmented.

Home communications in the future using optical fiber will integrate these various functions into an overall system, of which there will be communications, security, and computing subsystems, Corning believes.

As mentioned in Chapter 8, Bellcore researchers Paul Shumate and Leslie Reith developed a laser packaging design in late

1986 that incorporates a housing from the inexpensive compact disk laser packages. Both researchers acknowledged that additional hurdles need to be cleared before the design is practical. But "the cost factor now looks very manageable," Reith observed. "We're hopeful of seeing lasers in the local loop a lot sooner than previously expected."

Lucky downplayed competition between Bell Labs and BellCore, noting that neither BellCore nor the Regional Holding Companies it works for can manufacture equipment. "We frankly respect their results and a lot of those people used to be our friends and work with us right here." He continued: "The competition is Japan. It's the world against Japan."

Yet even the Japanese were having problems manufacturing these sophisticated components en masse by the end of 1986, Lucky observed. "They plan an evolutionary approach where they build up production and build down the cost," he said. He characterized the process as trying to build a Volkswagen after having only built Rolls Royces.

It is no wonder that bringing fiber to the home and office is gaining widespread attention. By the 1990s, Shumate predicts there will be hundreds of thousands of installations annually, and that number will increase to millions of installations each year. At those rates, penetration along the lines of what is happening with cable television (50 to 60 million homes) is possible in the next 30 years.

Assuming an average distribution length of 2 km, 50 to 60 million installations means that there will be a market of more than 10^8 each of light sources and detectors, according to Shumate. There will also be the need for something between 10^9 and 10^{10} high-speed integrated circuits. "Compared with today's long-haul market, this is over 10 times more fiber and over one hundred times more devices," he said.[1]

An interesting product AT&T introduced in the spring of 1987

---

[1]*The 1988 Fiber Optics Sourcebook,* Phillips Publishing Inc., Potomac, Maryland.

in its attempts to hasten the fiber-to-the-home market is Fiber Ready, which contains both copper twisted pairs and optical fiber. The fiber "seed" in the cable allows a telephone company to install copper phone lines for service now and use the fiber for later applications, without further installation expense.

One of the major issues is whether to pursue the LED-multi-mode fiber solution (characterized by Lucky as the "low tech" answer) or lasers and single-mode fiber (labeled the "high tech" solution). Making no doubt about his own leanings, Lucky noted that "if someone put a multi-mode fiber into my home, I'd be offended."

Yet there are still people at AT&T pushing for the LED-multi-mode scenario, despite what has happened in the past. "A lot of people in our company work in this field and believe the right solution is the cheap solution, that you do not use lasers and you do use multi-mode fiber. Yet they are told over and over by customers that they do not want that, they want the stuff that lives into the future," says Lucky.

A strong consensus among those at OFC '87 preferred single-mode fiber as the medium of choice for the subscriber loop. This was initially mentioned by plenary session speaker G. Heydt of Heinrich-Hertz Institute in West Germany, and was reinforced by BellCore's Peter Kaiser and others.

"Because of their superior transmission characteristics, single-mode fibers are the transport medium of choice for future broadband-ISDN systems," Kaiser said. "The use of single-mode fibers with their nearly unlimited bandwidth reduces the likelihood of premature obsolescence of the installed fiber plant and enables the definition of BISDN (broadband ISDN) interface rates and channel structures, which are based on close-to-optimum coding rates for extended quality and high definition television."

A central reason for using single-mode is to accommodate the high transmission rates that companies including BellCore and Fujitsu believe will be possible in the subscriber loop. Single-mode fiber is also expected to accelerate the advent of coherent transmission and integrated optics, as well as more sophisticated lasers.

Again recalling the debate of a decade earlier, there was disagreement as to whether lasers or LEDs would be used. This was an area of heavy dispute at OFC/IOOC '87 and promises to result in major disagreement as fiber comes into the loop.

Lucky sagely points out that—while you can change the electronics—"when you bury fiber you're going to have to live with that for a lot of years because that depreciates over a lot of years."

However the scenario is played out, OFC will continue to be the major forum. The conference has grown with the industry and the exhibit area has undergone a corresponding growth.

Certainly another important factor in getting the economics to fit is making equipment that will not result in large labor expenses. ITT Cannon has developed a rapid termination process, called fiber-lens fusing, that the company believes will significantly reduce the labor time and cost required to terminate optical fiber. Cost reductions of up to 79% are predicted by Les Borsuk, director of new products, fiber optic technology, for the Microtech Division of ITT Cannon.

The it-won't-happen-tomorrow-so-let's-not-kid-ourselves attitude that Lucky and Bell Laboratories have adopted with regard to bringing fiber to the home is probably very realistic. He acknowledges that AT&T has "backed off a bit," that it has come to the realization that the only way it is going to happen from their perspective is through voice telephone. Most homeowners are not going to go out of their way to pay extra money for services they don't know much about.

The cable television industry seemingly would have a golden opportunity to provide these services as add-ons to entertainment television in the meantime, but they have not decided to do so. "It's a very flaky business industry and they have not been particularly helpful," says Lucky.

Southern Bell is using multi-mode fibers in its pioneering Hunter's Creek project in Orange County, Florida, to extend fiber to a residence's garage, where optical signals will be changed into electronic signals. The two multi-mode fibers initially can provide 36 channels, potentially expandable to 54 channels.

While acknowledging that Southern Bell is spending lots of money on the trial and noting that it is more expensive than copper at this point, Southern Bell's Karen Mangum also notes that "we're looking for the crossover point where it will be economical. We know we are going to get there." Mangum is staff manager, marketing technical support. A related goal for Southern Bell in the Hunter's Creek project is to increase the transmission rate from 45 Mbps. (See Figure 1.)

One can better understand the potential for fiber optics with

Figure 1. A future fiber optic loop system could feed digital cable television signals over fiber to private homes, where a decoder in the garage could supply analog electronic signals to an AT&T channel selector atop the television set. (Photo by David Pardee; courtesy AT&T.)

Lucky's observation that photonics is in its "primitive" stages of development compared to electronics.

That does not mean that there will continue to be new generations of fiber optic systems. Stewart Miller has observed that "in a technological sense the major challenges that make something possible have been met."

And Richard Dixon, director of Bell Labs lightwave devices laboratory, feels strongly that the current generation of fiber optic technology will continue to serve marketplace requirements. "The silica based fiber and the present generation of components hung on the ends of it will last a long time," he believes.[2]

What Lucky's observation does mean is that optics will continue to play a more prominent role in the interaction of optics and electronics that has come about. In an article subtitled "In fiber optics, a burst of light is worth a million words," Lucky refers to these changes as coming about by improvement of the "plumbing" modules. These are amplifiers, switches, couplers, filters, and isolators that can be integrated into microcircuit chips.[3]

"Commercial optical fiber systems now generally transmit data on a single wavelength, like a single radio station," Lucky writes. "The enormous bandwidth of the optical fiber remains unexploited because of difficulty in establishing and manipulating subchannels."

While Lucky acknowledges that progress through late 1986 was slow, he believes that breakthroughs can come none too soon. "The current flow of information is growing too large for electronic processing," he wrote. "Our useful electron will be replaced by the upstart photon."[4]

Fiber optics has already begun to blur the conventional tele-

---

[2]"Longhaul Consolidation: Further Dancing Around The Monolith?" *Fiber Optics News,* January 27, 1986.
[3]"Message By Light Wave," by Robert Lucky, *Science '85,* November 1985.
[4]Ibid.

phone distinction of loop and trunk transmission networks, says Robert Stickle, who has served as a consultant to AT&T. It's all "merging into a single, seamless whole," he has noted. High-capacity fiber optic trunking, for example, may be used in an office building. "Even the smallest system may be flexible enough to carry voice, video, and data in loop and trunk applications," he said.

The result will be "a flexible, expandable arsenal of communications equipment for network designers, economics of scale and fast response time for service providers, and new economic services for the customer—all at the same time," says Stickle.

Such systems would take advantage of even better electronics being developed. AT&T, for example, has an intensive program at Bell Labs attempting to develop systems operating at 6.8 Gbps. Such capacity is required to satisfy fiber-to-the-home requirements, according to Lucky.

The impact of these networks of tomorrow will be profound. Dr. Kao analogizes the networks to new interstate highways attracting a host of new businesses around them.

Lucky uses a similar analogy, that of "a superhighway through undeveloped land." This superhighway, he says, will "undoubtedly draw new businesses, commuters, and even Sunday drivers."[5]

The technologists such as Kao, Miller, and Schultz so instrumental in birthing and raising fiber optics can only guess along with the rest of us what wonders have been wrought. Perhaps this is as it should be, since as scientists they live somewhat apart from the societal impacts of their actions. Perhaps accustomed to controlling the variables that have made fiber optics what it is, these scientists cannot really control how fiber optics will be employed. They can only hope that its impact will be positive and that it will be put to good use.

The success fiber optics has enjoyed has come from the fact

[5]Ibid.

that it has been in accord with nature, explained Alan Chynoweth at the OFC '85 conference. A fiber optics pioneer in his own right who has never lost track of the human element involved, Chynoweth advised that fiber has been successful and would remain successful—regardless of what proponents of other technologies said—because of its ability to work within the boundaries of nature's laws. Whatever additional success the technology has will only come about if it stays within those boundaries.

Chynoweth recalled the tale of King Canute, the tenth-century Danish king of England, who was so taken by the flattery of his courtiers that he sat out by the ocean in an attempt to stem the ocean's tide—and failed miserably.

"My confidence in the irresistible advance of this technology is based on something much more sound than mere bravado," Chynoweth proclaimed. "It is because technology advances to the beat of a different drummer. Technology advances based on nature's laws, and in any confrontation between nature's laws and the man-made laws of latter day King Canutes, the advantage is always ultimately on the side of nature."

Chynoweth and other scientists believe a major use of the extra bandwidth will be for imaging to the home and office. "One need in particular that stands out in this coming age of electronic libraries and other information bases is the ability to browse or scan in much the same way a person scans a newspaper or browses through a book," he said.

"These networks will facilitate new services that customers will expect the telecommunications companies to provide, services such as ready access to a wide variety of information, education, and entertainment sources and to libraries for archived knowledge—including literature, music, plays, and video programs on demand," according to Chynoweth.

Also accessible, according to Chynoweth, will be topical information regarding weather, sports, catalog shopping, electronic mail, and financial transactions. "The possibilities are limited only by human ingenuity and invention of new services," he said.

For his part, Kao would like to see a data bank service, similar to the type of services that Merrill-Lynch provides for the stock market, "but in much grander scale." Such a service would involve "computer-aided design with automatic ordering of selection of equipment, which would be built in the data bank so that it would directly impact the productivity per capita in producing it."

The networks of tomorrow will provide something that will be a first for mankind, cheap bandwidth. Kao characterizes fiber as "a zero cost to the infinite bandwidth zero loss medium," meaning essentially that we will have plenty of something that was once held to be costly.

The result is likely to produce some societal confusion. "It's like when you have access to a water supply once every four days for three hours, on rations, and suddenly you can use it without restrictions," Kao says. "You are a little unsure what to do with it."

Despite the prospect for incredible amounts of bandwidth, Lucky believes we will find ways to use it, even if much of that involves learning how to squander it.

People in New York City were once limited to a gallon of water daily; now they use "fantastic amounts" because it's easy to get, says Lucky. "It's not that you intentionally want to waste it, it's that it is a convenience." And Kao, ever the seeker and optimist, believes: "We will use it intelligently and very beneficially."

A question that inevitably arises is where the technologists such as Kao, Miller, and Schultz end up once the rewiring of America has been accomplished. The question may be moot, at least for the time being, because challenges continue to arise.

One such challenge, for example, is the coming of the all-photonics network, when photons will speed the flow of information through society, unencumbered by the slower electrons. Photonics in the twenty-first century will be analogous to what electrons represent this century, says Robert Sprinrad, director of systems technology at Xerox Corporation.

Twenty years after he first told the world that a technology known as fiber optics was going to work, Kao headed up a dedicated team of ITT scientists attempting to transmit terabits of information for the further utilization of the potential of fiber optics. He returned to Hong Kong to serve as vice chancellor of Chinese University in autumn 1987. Kao founded the electrical engineering department there in the early 1970s.

Sixteen years after he and two associates found the low-loss glass to help prove Kao right, Schultz has received a patent for a hermetic coating that could make fiber safe for undersea applications. The company where he worked until 1986, SpecTran Corporation, hopes the technique will help bail it out of a financial crisis. Schultz has since moved over to Galileo, one of the first U.S. companies to recognize the promise of fiber optics back in the 1970s.

Schultz and the other two members of the original Corning team that created the first low-loss fiber were duly represented and recognized at OFC/IOOC '87. Schultz served as technical program cochairman. Robert Maurer received the first John Tyndall Award, presented to him for the initial efforts. Donald Keck serves on the Optical Fiber Steering Committee for OFC.

More than a decade after he spearheaded the first OFC conference, Miller is still a regular attendee and active in the fiber optics community. Miller has observed that fiber optics will "go into practical use in all parts of our lives," but that most of what happens will be out of sight of the user. The end result, then, would be that for most there will be a better quality of life with little understanding of how it came about.

In decades to come, perhaps the realization that America was rewired with fiber optics will only come by realizing what came before it. This may result from viewing a photograph or observing in a museum a large grotesque ball of copper cable that was an example of the way people communicated before the advent of a revolutionary technology known as fiber optics.

Index